Politics as if
Evolution Mattered

Politics as if Evolution Mattered

Darwin, Ecology and Social Justice

Lorna Salzman

iUniverse, Inc.
Bloomington

POLITICS AS IF EVOLUTION MATTERED
Darwin, Ecology and Social Justice

iUniverse books may be ordered through booksellers or by contacting:

iUniverse
1663 Liberty Drive
Bloomington, IN 47403
www.iuniverse.com
1-800-Authors (1-800-288-4677)

ISBN: 978-1-4620-3475-8 (sc)
ISBN: 978-1-4620-3476-5 (hc)
ISBN: 978-1-4620-3477-2 (ebk)

Library of Congress Control Number: 2011914822

Printed in the United States of America

iUniverse rev. date: 08/22/2011

Contents

Introduction

Evolution, more than any other scientific field of study, intersects with almost every social and political subject today. Daniel Dennett has described this characteristic felicitously by calling it the "universal solvent". It has the capability of shaping public discourse in a more rational and ethical direction than the mass media, pop trends and political ideologies allow.

Activists and the general public might be puzzled by the notion that an understanding of evolution could enrich and increase the effectiveness of movements for social change. While these movements (after several decades of foot-dragging) are slowly absorbing the ecological principles at work in many of today's challenges such as climate change, nuclear power, genetic manipulation and biodiversity losses, most people have not yet grasped the connection of ecology and evolution. Once grasped, this understanding will have tsunami-strength impacts on the strategies and objectives of these movements.

Progressive social change movements in particular are susceptible to the cry that "the personal is political" and give weighted attention to trends such as Identity Politics, spirituality and the proclamations of religious leaders who have finally begun integrating environmental concerns into their sermons—and unfortunately vice versa, as did one leading liberal rabbi who called for a president "filled with the spirit of God".

Add this to the body of post-modern thought which regards science as "socially constructed" and denigrates the notion of universal truths, and one starts to understand how irrationality, distrust of science and a belief that subjective thought reflects reality actually represent regressive, even subversive, societal trends that undermine science, education, social justice, ethics, and our political process.

New challenges to the direction and fate of civilization arise each day. Environmental articles and books document the degradation of the earth's natural systems and the impact on human welfare. But the response of the environmental movement and social change activists remains sorely inadequate. Consequently progressives regress to simple moral issues like peace and poverty rather than confront the complexities of science and the challenge of reconciling innate human imperfection with democracy, peace and human rights.

While scientists and environmentalists try to understand the relationships of non-human animals to their environment, they have neglected to do so with humans and human society. Unlike nonhuman species, humans are gifted with the ability to choose with deliberation the conditions of their existence. Tragically, their choices in contemporary industrial society, even when motivated by ethics and compassion, are proving to be not merely inadequate but fatally wrong.

It is perhaps human nature to resist the stark reality of fallibility and the role of chance in evolutionary processes. But it is precisely the lack of an evolutionary perspective in movements for social change—a recognition of the "laws of nature"—which deprives them of the necessary insight into the parameters of human existence, the insight that could provide a substantive guide for a social reform agenda not in conflict with the natural world. A growing backlash against science on the Left and the Right only exacerbates the situation.

The central dilemma of modern humanity is the failure to adapt our behavior, institutions and objectives to certain realities which we ignore or defy through our faith in technology and religion. This defiance is at the root of our ecological crisis along with the delusion that the human species or society can be perfected.

There are four areas most lacking in ecological and evolutionary understanding, broadly speaking: Ideology (religious or secular); Technology; Social Justice; Human Ethics. Under Ideology are grouped

2

movements such as New Age thought, Marxism/socialism, faith and spirituality. Under Technology there are genetic manipulation, nuclear technology, resource exploitation and energy production. Under Social Justice and Ethics we have equality, equity, human rights, class/race/ gender issues, intergroup aggression and militarism.

Much effort is expended on influencing and reforming human behavior and cultural choices as if these were fresh clay to be molded at the whim of whomever is in charge. Since Earth Day 1970, when the word "ecology" emerged as a potential organizing idea, activist groups have competed with each other to define *The Problem* and, not surprisingly, have come up with solutions that fit their own political model, much like the proverbial three blind men groping an elephant and postulating three different shapes. Some activists postulate capitalism, others population growth, others human greed, others corporations, others the male gender, and others the wrath of one god or another. Few regard humanity as being one of many products of evolution or a species not endowed with special rights.

This book lays out the "evolutionary imperative" which, upon closer examination, reinforces not what some consider inherent human evil ("nature red in tooth and claw", competition, domination) but concepts considered ethical today: non-hierarchy, functional equality, cooperation, and interdependence. The failure of progressives and the Left to understand evolution's lessons and the failure of liberal movements to underpin their political goals with science and rationality rather than abstract moral precepts or *a priori* doctrines, or to place humans within the evolutionary chronology, has created a large void into which rabid ideologues of all stripes have intruded with their own arbitrary and capricious dictates and who now dominate public discourse. The centrality of evolution and ecology—two sides of the same coin—must be recognized before any cohesive, consistent social justice movement can succeed.

Neither Nature nor evolution can provide a moral guide for human behavior or technological and ethical choices. What they can provide is an impartial, scientific explanation of which choices are most likely to enhance human welfare and survival, and which ones are more conducive to societal collapse or species extinction. The burden of decision-making ultimately rests on human intellect and rationality, with the hope that

3

humans will alter their behavior and institutions in the right direction. This collection of essays will try to explain the relevance of evolutionary thought to the pressing ideological, technological and ethical debates of our time.

Ω

Why Darwin Matters Today

"... it took nature more than a billion years to develop a good worm."

The Sea, Robert C. Miller

"Miracle . . . is simply a synonym for the total absence of explanation."

River Out of Eden, Richard Dawkins

"... the framework of human life is all life . . . it is life, not man, which is the main contour, and it is ecology in general where human ecology is to be found."

The Subversive Science, Paul Shepard and Daniel McKinley

The theories and writings of Charles Darwin, in particular his *On the Origin of Species*, have grown into the foundation of all modern biological thought. It is no exaggeration to say that the disciplines of biology, ecology, modern medicine and agriculture, among others, could not exist without Darwin. Without him, these would be forced to rely on random observation and anecdotal experience, mere trial and error, with all the inherent handicaps.

Biological and medical science would operate in a vacuum, accumulating clinical and experimental evidence that would lack coherence and predictability. The development of modern medicines and the refinement of food crops (conducted by traditional farmers over millennia through simple observation and experiment) would be haltingly slow. Humanity would have virtually no understanding of biological systems, their functions or their importance. The treatment of disease would be limited to treatment of symptoms alone. Management of natural resources would lack a scientific basis. In truth we would still be in a Dark Age of science, scarcely any better than the shamanism practiced in some parts of the world (which as a cultural practice has some social validity in non-industrial societies). As Theodosius Dobzhansky cogently observed, nothing in biology makes sense except in the light of evolution.

No original aspect of Darwin's theories on evolution and natural selection (and evolution is accepted as fact today) has been proven wrong, with the exception of Darwin's belief in inheritance of acquired characteristics, an understandable error given the total ignorance of heredity mechanisms in his time. The basic precepts of evolutionary change and natural selection remain the foundation—the working hypotheses—of scientific research and experiment. No one has yet proposed any different working hypothesis, nor has any experiment indicated that the use of these is flawed or erroneous. In every laboratory and field study, analysis and re-analysis, evolution and natural selection have proven to be unshakeable and valid. The reliance of biology on the fact of evolution and the theory of natural selection is as justified as the reliance of engineers on mathematics and physics for building bridges, skyscrapers, airplanes and machinery.

Even the claim sometimes made that natural selection cannot be demonstrated or proven is false. Chemical manufacturers constantly reformulate pesticides and pharmaceuticals because of the spread of resistant microbe strains or populations of insects, most notably the AIDS virus. Every year the Centers for Disease Control, as well as drug companies, must come up with new drugs as resistance to earlier ones grows. Malaria, the largest disease killer on earth, has resurfaced with a newer deadlier strain that is resistant to nearly all existing drugs and which can only be treated by a dangerously toxic drug that itself can be lethal (and which itself will eventually become useless). Influenza, which killed at least 40 million people in the 1918 global epidemic, must be confronted every year by scientists who try to predict what type will appear so that they can design the proper vaccine. Recent studies by two scientists, Peter and Rosemary Grant, of the finches in the Galapagos Islands—the same ones observed by Darwin—have shown that evolution by natural selection is actually observable over short time periods (see *The Beak of the Finch* by Jonathan Weiner, Vintage Books 1995).

While no reputable biologist doubts or uses anything but Darwinian precepts for research, the philosophical and ethical implications still trouble many people, and not just those who adhere to fundamentalist religions. Some of these concerns are over the implications of evolution for religious believers in a deity; others have difficulty with the notion that the human species is not intrinsically "superior" to other animals; yet others are disturbed by what they see as violence and competition in Nature and wonder whether this is inherent in humans; there are those who refuse to accept the role of chance and stubbornly adhere to belief in purposeful design; and there are those who may believe in evolution but insist on the notion of "progress" and "purpose" in Nature.

There are many reasons for this: a general lack of familiarity with science and how it functions; an expectation that science can attain "final" answers; an inability to regard science as a process, not a religion; the prevalence of a broad spectrum of differences in interpretation of scientific data; the unavoidable ideological bents and biases of individual scientists; a reprehensible emphasis by the mass media on discrete facts and events rather than on processes or systems—all these, plus the distortions of self-interested political and religious ideologues who hope to promote their own agenda, have combined to make science in general and Charles Darwin in particular suspect.

One would have thought by now that rationality, education and evidence would have dispelled the doubters. But apparently human nature still has a tendency to pick and choose the facts it wants to believe. No one wants to subject her heartfelt prejudices to an impartial test without being certain that the result will be in her favor. Far safer and more comfortable it is, therefore, to avoid such tests. And far more comforting it is to find other "evidence" that will substantiate your beliefs.

While the religious creationists and supporters of "intelligent design" have done this with a vengeance, they are not the only culpable ones. Irrational beliefs, cults, and ideologies have always existed, but for some reason the past four decades have been particularly rife with these, culminating in the appearance of New Age groups with distinctly pre-Enlightenment views, whose supporters, far from being uneducated or uninformed, come primarily from relatively affluent educated circles. There is also an

7

incredible distrust of science and scientists in general, due to a lack of education in many cases but also because there is no denying the abuses that have been committed by science and technology, such as nuclear energy, toxic chemicals and genetic manipulation. A favorite simile of journalists writing about social and economic issues is the use of the term "Darwinian struggle" or something like it. Put these all together and you have a large body of citizens who can easily be misled into thinking that Darwin is responsible for it all.

There is also the fact of serious disagreement by credible scientists on the intricate details of how evolution and natural selection work, compounded by the large time scale involved in these processes. There are debates on the importance of adaptation, on the target of natural selection, on the issue of gradual vs. rapid evolutionary change, on the definition of species, and so on. But despite the acrimony that sometimes accompanies these arguments, contrary to appearances and inflammatory media pieces, none of these competing hypotheses in any way undercuts or challenges the acceptance of evolution and natural selection.

Several things should be kept in mind. First, nothing was known in Darwin's day about genetics and heredity; it was not until the early 20th century that the findings about these subjects, based on Gregor Mendel's experiments, were added on to the concepts of evolution and natural selection, to form the "Modern Synthesis". This synthesis enabled scientists to refine and expand their understanding of how evolution works. To blame Darwin for shortcomings in his theories when the science was lacking to even ask the right questions is obviously unfair. For example, Darwin assumed new species only arose out of one pre-existing lineage. Since then additional theories on speciation have arisen of which Darwin never dreamed, most notably speciation arising from geographic isolation of populations within a species, and the subsequent development of isolating mechanisms that arise and keep these new species separate from earlier ones. A whole new field of theory and study has arisen around the hot issue of what defines a species. But this only builds upon rather than replaces Darwin's theories.

Second, these essays do not discuss sexual selection, another aspect of evolution which involves the selection by females of males of a species who are visually or behaviorally most attractive to them, such as those with conspicuous or aesthetically pleasing characteristics or behavior, such as the peacock's large ornate train or the spectacular breeding displays of prairie chickens and Birds of Paradise. "Sexual selection is . . . one of the most potent causes of speciation in animals . . .", says evolutionary biologist Douglas Futuyma.

Third, the subject of sociobiology, or what is now called evolutionary psychology by social scientists, has become one of the touchiest subjects of our time, though none the less fascinating and important. Initially it was called ethology and started with the work of Konrad Lorenz and Niko Tinbergen and their field observations of animal behavior. It later evolved through the writings of, among others, Edward O. Wilson, who later expanded his theories about the simultaneous evolution of human culture and the human species. On the way it was tainted (unfairly) by charges that it supported the "biology is destiny/genes determine human behavior" concepts, pop media notions that no respectable scientists, Wilson included, have ever claimed. Finally it has at least become a field of study worthy of pursuit, though tinged, according to some on the Left, with racism, fascism and sexism.

The "Nature vs. Nurture" debate still rears its tiresome head, and when it does it needs to be dealt with, especially in light of the inability of the general public to examine scientific controversies in a politically neutral manner. Few if any scientists deny that there is some genetic basis for human behavior as there is for human appearances (phenotypes). It could hardly be otherwise, given the fact that the human brain, source of culture and ethics, evolved along with all our other organs and systems. How and how much heredity contributes to human behavior, values and culture, whether we should continue investigating and what we should do about it, remain the subjects of raging ideological battles, which are often portrayed as political (Left vs. Right), but which really reflect different attitudes about the criteria for ethical decision making. Thus these debates should be removed from the scientific arena and resolved in the social arena.

In fact one of the reasons that Darwin withheld publication of *Origin of Species* for so long was not because he feared antipathy to the notion of evolution from earlier life forms—this notion had long been extant—but because his theory implied that the human species and its individuals arose from a material basis. This materialism of course infers that the human brain evolved in a similar manner to that of other organs, and that humans similarly evolved from other animals. The only conclusion to be drawn from this fact is that if the brain has a material basis, so does the human mind, including religious beliefs. And this in turn leads to the inescapable conclusion that the notion of a deity is simply a product of the human mind. In this light the present-day arguments of creationists and "intelligent design" proponents can readily be seen as feeble attempts not only to refute evolution and impose irrational religious beliefs on everyone else, but ultimately to rebut materialism *per se.*

A clearer separation is needed for rational political discourse, one that understands that human ethics, while highly likely to have a strong evolutionary base (see Michael Ruse's writings), exist in a context of free will and choice, and thus are more appropriately raised in a discussion of human social and political change rather than in a discussion of human origins and genetics. Moreover, no aspect of genetically based human behavior is fixed or compelling, including those aspects of adaptive human behavior that we voluntarily accept, such as incest avoidance. Such taboos are accepted not because we consciously know they are genetically harmful but because such ethical choices are defined in abstract terms called epigenetic rules and are not articulated in biological terms. We don't need to know that incest is a bad biological idea; we just dislike the notion and that is enough, in most cases, to discourage it.

Indeed, epigenetic rules are really the most effective method for preventing counter-adaptive behavior because they function perfectly in the absence of scientific evidence or intellectual comprehension. That natural selection enabled us to develop these rules is almost by definition a proof of natural selection, because early humans lacked sufficiently advanced cognition of what constitutes non-adaptive behavior. Early humans knew nothing about the genetic dangers of inbreeding; instead, those early humans who were not repelled by sexual relations with family members had less reproductive success and failed to produce more offspring with such tendencies than those who were instinctively resistant to such relationships. The latter eventually out-reproduced the former, allowing the instinctive avoidance (the epigenetic rule) to spread widely in the population. Our contemporary abhorrence of incest has its roots in our ancestors' distaste, upon which natural selection conferred an enhanced reproductive edge.

In some of these political exchanges, those who, for want of a better category, come from the Left traditionally condemn the notion that human behavior may in some measure be biologically "determined" and go to great lengths to discredit scientists like Edward O. Wilson and his theories about the biological basis of human culture. Aside from the fact that neither Wilson nor anyone else argues for biology as a supreme authority in human society—Wilson argues more for respect and deference—and that humans, like many animals, can make behavioral and normative choices (as well as mistakes and misjudgments), those who denigrate even the study of the evolutionary roots of human behavior ignore the most important aspect of evolutionary theory: its lack of determinacy and hierarchy.

If nothing else, the work of Charles Darwin and his successors in the field of genetics has clearly shown us that there is no biological basis for the concept of socially repressive dogmas or discrimination on the basis of physical characteristics. Similarly, evolutionary theory rebuts the notion of any non-material origin of the human species. Evolutionary theory thus carries within it—in fact made it possible to define—the notion of biocentrism (functional equality of all living species) and the corollary absence of any biological basis for hierarchy in human social, political and economic relations. Far more effective it would be, for the Left, to focus its ire on religion, which has a much longer track record than biology in oppressing humanity.

Finally, a word about spirituality is needed. There is no intent in this book to ridicule individual spiritual beliefs or dismiss private religious practice as motivating factors for those who support ecological causes. The notion that a political movement must be based on a purity and uniformity of intent and motive is anathema to the concepts of cultural diversity, cooperation and mutual respect.

Organized New Age movements and groups, some of which border on cults, bear no more resemblance to individual spirituality, ethics or religious beliefs than organized religion does. By definition spirituality is a subjective, private condition that may motivate some individuals as much as political issues motivate others. The New Age beliefs cited herein are suspect not because they are spiritual but because such beliefs, like any other, cannot be the basis for a political or ecological movement, much less for scientific research. Were the environmental or Green movements to make this error they would end up as religious rather than ecological movements, a Green "Moral Majority" of fundamentalist authoritarians in serious contradiction to their own principles.

Those with strong spiritual or moral beliefs, such as animal rights activists, who wish to have their opinions taken seriously or contribute to social change need to present persuasive rational and *ecological* arguments to make their case, in much the same way that marine mammal protectors have done. In their own sphere of influence they have the freedom to pursue their objectives as they prefer but in the broader societal sphere they cannot expect their personal ethics to prevail without stronger, science-based evidence.

Rehabilitating Darwin is a large task and one best left to evolutionary biologists in the boxing arena with creationists. Rehabilitating an understanding of evolution as the other side of the ecology coin, however, has practical implications for environmental and social justice activists, who sorely need intellectual rather than ideological weapons to turn back against corporate greenwashers and biased media. Ecological principles underlie ecological relationships, and practically every environmental battle revolves around such relationships, their character, their value and the threats thereto.

Environmentalists who are accused of being merely obstructionists, fearmongers and radicals with hidden agendas can dispel such charges only when they understand and utilize the tools provided them by the scientifically impartial concepts of ecology and evolution. A battle, not to mention a movement, founded on objective reality, on "real life" as it were, is not so easily lost as one founded only on abstract or subjective credos.

Ω

The Evolutionary Debate

In the early 1990s, my husband and I were visiting our daughter in Brighton, England, for Christmas and decided to visit Down House, in Downe, Sussex County, England, Charles Darwin's home for the last forty years of his life, where he wrote, among other things, *The Origin of Species*. We drove north towards London to a typical middle-class suburban town, turned off the main street and drove along a tree-lined road that looked like something out of a suburban American neighborhood. After about a mile the road abruptly turned to the left; it was now a narrow country road with open fields and farmlands on both sides. We proceeded for another mile or so, passing a large whitewashed house behind a tall wall, and as we did so, my memory jogged me as I recognized the house: "That's it!" I shouted. We doubled back, entered the gate, parked and proceeded to spend the next few hours in the hallowed halls and rooms where Darwin revolutionized the life sciences and scientific thought.

Except for the American biologist living there as resident director, and a pair of Japanese journalists doing research for a film, the house was empty of visitors. In one long room overlooking the large fields in the rear of the house, glass-topped cabinets lined the walls. They contained disparate artifacts, memorabilia and some correspondence relating to the Darwin family. As I browsed, one letter, quite legible, attracted my attention and I read it in full.

The letter, written shortly after Darwin's death in 1882, was from a member of the Downe landed gentry to a local merchant who had provided goods and services to his family for many generations (clocks if I remember correctly). In civil but firm tone, the writer informed the merchant that a mutual acquaintance had seen him attending the Westminster Abbey memorial service for Darwin, and therefore, due to the blasphemous and heretical nature of the beliefs expounded by Mr. Darwin, he was now forced to terminate their long-term commercial relationship.*

A quaint, amusing anecdote showing benighted Victorian attitudes, you say? Attitudes that surely are no longer prevalent in our contemporary age of rationalism and scientific accomplishments? Wrong on both counts. We are now facing the same encounters of the theological kind that Darwin faced in 1859.

Like irrationalism and superstition, certain themes and attitudes have recurred over the centuries in Western philosophy, notably the notion of progress in human endeavors, the purposefulness of life and the universe, and the attempt to detect order out of seeming chaos. Until Darwin's time, science and philosophy were united in natural theology, in which the existence of a creative deity and a grand plan was generally undoubted; even some of the most brilliant progressive scientists, including those whose writings hinted at evolutionary thought, hung on to their creationist beliefs, at least until the publication of *The Origin of Species*. More particularly, there was little challenge to the notion that humans were obviously at the pinnacle of life and that their existence was due to the plan of a creator.

Thus, it is not surprising that Darwin—who did not even deal with human evolution in *The Origin of Species* for fear of alarming his wife and concealed his emergent atheism as best he could—created such an intellectual ruckus. It was not that the idea of evolution of present life forms from earlier ones was totally new; indeed, it had been floating around for years, even centuries, waiting for a genius like Darwin to put it together in a theoretical fashion. Lamarck, for example, believed in evolution but espoused the theory of inheritance of acquired characteristics, which was an intellectual dead end. In fact Darwin supported this idea too, because no one, Darwin included, knew about the existence of genes. Darwin did, however, hold the idea of particulate inheritance even without knowing about Gregor Mendel's pea experiments. (Lamarck's and Erasmus Darwin's

---

* A re-visit ten years later to the gussied-up, over-renovated Down House revealed the dismaying removal of this letter to archives in Cambridge or London.

belief that an organism passes on newly acquired traits as adjustment to the external world really represents a primitive notion of what Charles Darwin later said: that creatures are products of both heredity and environment). Neither the term evolution nor its concept were original with Darwin, who called his theory "descent with modification": new life forms originating from older ones.

There is a famous anecdote about the two Victorian ladies taking tea together. One reports to the other the latest gossip: the absurd theory that humans are related to apes. To which the other replies: "My dear, let us pray it is not true. But if it is, let us hope it does not become generally known." Unfortunately for Bishop Ussher, a creationist who said the world was created in 6006 B.C., and for doubters like naturalist Louis Agassiz, the general knowledge was out. And being related to apes was to be the least of it as the world would find out soon enough.

To put it succinctly, biocentrism had just been invented and materialism reinforced, as well as the social consequences of Darwin's theories and writings, of which Darwin was fully cognizant. The grand plan of a deity, the immutable structure of the cosmos, the lofty perch of humans at the pinnacle with all their wisdom and power, their superiority over the lower creatures, the inevitability of and necessity for human existence—all gone. And also gone was any scientific justification for dominance and hierarchy in social relations and systems. (What is truly strange in our day is that radical social change movements have failed to understand the liberatory implications of evolution and ecology.) At one fell swoop, Darwin had swept away the philosophical and theological foundations of kings, emperors and popes, the superstitions of the fearful, the pride and arrogance of humankind regarding the rest of nature, the magic wand of medieval wizards, prophets and seers—all crumbled to dust by the wayside, gone forever. Or so it seemed.

Now let us move forward to our time and look at some contemporary attitudes towards not only Darwin but towards evolution in particular and science in general. These new voices have a familiar echo of those older voices that reviled Darwin, ridiculed his theories, took umbrage at his "blasphemy" and "heresy" and they are talking about "purposeful creation", "the special place of humans", "God's plan" and "evolutionary progress" again! These are not only the voices of religious fundamentalists, creationists, or promoters of "creation science" or "intelligent design" but, sorry to say, reputable biologists, physicists and socially progressive intellectuals.

A strange combination of awe and puzzlement over the complexity of the cosmos and the persistence of seemingly unanswerable questions raised by research in physics and cosmology have led to the unwarranted assumption that evolutionary theory cannot explain the "mystery of life" on earth and that only some kind of design or plan can be responsible. Evolutionary biologist Richard Dawkins calls this the "Argument from Personal Incredulity". What it reveals, however, is an abysmal understanding of the evolutionary process, namely the demonstration by Dawkins and many others that genetic variation and sufficient lengths of time quite suffice to explain how and why evolution works. The fact that doubters ignore or refuse to consider this evidence would seem to be *prima facie* proof that matters of faith have often pre-empted science, and that all arguments to persuade believers otherwise are doomed to failure.

Physicist Paul Davies said, in the *New York Times* (August 11, 1996), that the universe is not ruled by chance alone but by an "innate tendency to develop more complex structures . . . the universe now seems purposefully tailored to ensure the emergence of beings like us . . . or similar sentient creatures." In this same article, he states that he believes that God chose laws of nature that would guarantee the evolution of intelligent, self-reflecting beings.

Nicanor Perlas, agricultural scientist in the Philippines and winner of the UNEP Global 500 award, wrote in *Biopolitics* (Zed Press, 1995): "Biologists have evidence that non-physical, 'morphogenetic fields', not DNA, govern the emergence of form in living organisms...The past forms of organisms transmit influences to other organisms in the present and the future by means which transcend normal space-time conditions".

Edward Goldsmith, publisher of *The Ecologist*, in his book *The Way* (Shambala Publications, 1992), wrote: "Life processes at all levels of organization, including the evolutionary process itself, are designed according to the same plan, being purposeful, dynamic, creative, intelligent . . . the individual cannot conceivably be regarded as the sole unit of evolution; it can only be Gaia herself, and we can best refer to evolution as the Gaian process . . . The evidence for the purposefulness of life processes . . . is so great that its denial is inconceivable. Who could deny that the evolution of gills and fins by fish is purposeful to enable them to breathe and move about in their aquatic environment . . ."

Rupert Sheldrake, inventor of "morphogenetic fields" postulated to replace DNA as the force of evolution and inheritance, who was trained as a botanist, writes in *The Rebirth of Nature*: ". . . the vital force, spirit or energy flow can be only one of the aspects of life. The very fact that the same energy can take so many different forms means that something else must account for these forms themselves . . . (they) must owe their different characteristics to some formative principle over and above the flow of energy . . .".

All of these arguments are intent on denying randomness and indeterminacy any place in the evolutionary process, while replacing it with what seems to them to be the only logical alternative: a creative force acting with intent, foresight and design. Such alternatives, however, are neither credible nor rational and, like the creationist agendas being foisted on defenseless school boards, have no place in scientific discussions or education. Why not?

To answer this, we must enter into the field of philosophy of science if only superficially. First, it should be made clear that the philosophical discussions about evolution and the origins of life most often do not distinguish between the "how" and the "why", or between what has actually occurred as opposed to what might have occurred. Some Gaians point to the chemical ratio of atmospheric gases—most likely the only balance that allowed our form of life to appear and evolve—as being so finely tuned that chance alone could not have produced that precise ratio, and that therefore some larger force must have engineered the elements into the proper balance.

But even an amateur logician can see the fallacy in this proposition. If there were an infinite number of possible ratios, obviously any one of that number was as likely to prevail as any other. Had another prevailed, no life might have appeared at all, or perhaps an entirely different type. Even physicist James Lovelock, originator of the Gaia hypothesis, fell into a similar trap of illogic. In his *The Ages of Gaia* (W.W. Norton, 1988), he wrote: ". . . imagine that some cosmic chef takes all the ingredients of the present Earth as atoms, mixes them, and lets them stand. The probability that those atoms would combine into the molecules that make up our living Earth is zero. The mixture would always react chemically to form a dead planet like Mars or Venus."

But the original gaseous elements out of which the planets in our solar system formed were already in the form of molecules, not atoms (*Universe, Earth and Atom: The Story of Physics*, Alan E. Nourse, Harper & Row, 1969), and even Lovelock later came to understand this, when he later said: ". . . the early Earth is thought to have had on its surface the chemical compounds that are called 'organic'—such as amino acids . . .". In any case the argument about whether life could have appeared under different circumstances is the wrong question to ask; it is a truism that present life forms are adapted to the Earth's environment, and therefore that environment is the only one in which we can survive. But this in no way suggests that the present environment and life forms were the only possible ones.

Ω

Science vs. New Age-ism

Scientific hypotheses, in order to be credible, are required to be testable by other scientists. The Austro-British philosopher Karl Popper devised a criterion for separating out scientific hypotheses from unscientific ones (that is, separating out those which should be taken seriously from those not requiring further attention). It is called the falsifiability test. Simply put, a hypothesis should be taken seriously only if a test exists or can be devised that could, if carried out, demonstrate the falsity of the hypothesis. The previously mentioned hypotheses regarding purposefulness, plans, intent, and the non-evolutionary explanations for the existence of life forms are irrational because they fail the falsifiability criterion.

This has nothing to do with the *ultimate* truth or falsity of the hypothesis; the statement could theoretically be proven true or false by future advances in science or the discovery of new evidence. What counts is its testability; if a method exists that *could* prove the hypothesis false, then that hypothesis is considered a rational assertion worthy of further discussion or experimentation. By such standards, none of the quotations of Davies, Perlas, Sheldrake and Goldsmith are testable and they therefore lie outside the realm of scientific discourse. They belong to religion, spirituality, parapsychology, superstition or some other realm of belief because they cannot be tested by any means presently available to us.

Today we are faced with growing irrationality of the same variety that has plagued societies in the past, whether it be the Middle Ages burning of witches to counteract the Black Plague, the propitiation of gods and goddesses by ancient cultures, the religious hysteria that frequently consumes individuals and communities as they marvel at the liquefaction of Christ's blood or his miraculous image on a piece of linen, human sacrifices, seances, psychic communication with the dead, palm and crystal readings, astrology . . . and so on in perpetuity.

It is probably fair to say that such irrationality is deeply ingrained in the human psyche, perhaps even a biological by-product of human brain evolution. In evolutionary terms it may have had very real adaptive value, either biologically or culturally. Major organized religions are founded on a belief in imaginary gods and goddesses; and while we may deplore the authoritarian nature of some of these, it is likely that orthodoxy, dogma and hierarchical structure provided (and provide today) not only spiritual guidance but social stability in periods of stress and social chaos. I once read a thoroughly convincing explanation by a physician of why we often visually translate ordinary objects into threatening creatures. He postulated that for primitive humans living in wild nature with large dangerous animals lurking, it was far safer—and adaptive—for the human brain to turn a harmless dark rock in the distance into a bloodthirsty tiger, because, upon seeing the "tiger", the human ran away and lived to produce descendants with similar wariness. But for the human who mistook a real tiger for a dark rock, there was no future, and he or she left no descendants.

Primitive animist religions very likely have strong ecological foundations in that adaptive social practices are sanctioned and non-adaptive ones prohibited by taboos. The incest taboo is of course the most renowned one, as are, in general, codes governing marriage, sexual relations, property and family and of course food, all of which are closely related to survival and reproduction. We readily accept the notion of artistic creativity and genius, where visual and verbal artists produce startling original works out of their imagination (someone like Jorge Luis Borges comes to mind). We should therefore not be surprised at the human ability to conjure elves, fairies, genies, witches and banshees.

Now, science performs such explanatory functions, though without any attached moral component—a component that is or should be provided in any case by communities and secular society. But religions as well as New Age leaders and organizations continually try to place their imprint on secular movements and institutions, using the argument that pure rationality and science cannot provide moral examples or leadership. While organized religions will always be with us, the New Age movement has the potential to expand infinitely, pleading that it is giving voice and shape to human emotions that are expressions of some universal truths, while technically avoiding the vise of narrow or coercive organized religion. That the high priestesses of the New Age are playing the same role as popes and priests does not seem to bother many adherents.

Such New Age appeals to intuition, starkly anti-intellectual, will always attract followers because of their non-judgmental validation of any and all individual beliefs, a validation that, in common with religion, does not require any kind of objective "proof" at its foundation. They play on not only the dislike of coercive religion and its habit of showing moral disapproval of human behavior, but on the ubiquitous anger with the hard proofs demanded by science and, more particularly, the often uncomfortable conclusions that can be drawn from such proofs.

New Age spirituality is by definition comfortable, approving, welcoming. It makes no demands on an individual such as those of science or, more aptly, organized religion. It does not belittle or disparage emotions or irrationality. It does not pass judgment or ask embarrassing questions. Thus, it ultimately rationalizes and blesses each and every private and internal realm. New Age leaders have therefore moved in on religion in a big way, and in so doing can make the claim that they are not preaching religion but a different way of seeing the universe. But in some ways, with regard to civic involvement and social activism, New Age spirituality may be more divisive than religion.

Spirituality differs from religion in that it is private and particular to the individual, whereas religion involves a shared set of beliefs and values in the service of which the individual must set aside his or her personal morality. Whereas traditional organized religions bring assorted individual beliefs together within an historic collective body of thought or dogma, New Age thought starts with no such dogma and reaches outward to millions of individual beliefs to validate each and every one. Its appeal is therefore potentially far greater than that of organized religion. But the end result is the same: the recruitment of individuals into an overwhelming security blanket that effectively rewards the individual by validating the irrational and unverifiable. And by implication it denigrates or circumvents reason and science.

Anti-technology movements continue to appear down through the ages, and, notwithstanding scientists to the contrary, they more often than not have their foundation in well-founded rational fears; the 18th and 19th century Luddites feared, quite rightly, the loss of livelihood from machinery of the Industrial Revolution. Today, again quite rightly, we fear not the vengeance of gods but nuclear meltdowns, airplane crashes, chemical toxins, genetic manipulation, food poisoning, and so forth. In fact, it is fair to say that the scientists' faith in the social and environmental benignity of such technologies is far more irrational than the fears of the modern layman, given the very real consequences of such technological failure.

What is new, however, is the reaction against not just dangerous or untried technologies but against science as a discipline, as an intellectual endeavor and as a profession. We should not be surprised that religious fundamentalists and creationists would reject the findings of those whose research seriously challenges their religious beliefs by providing evidence about the origins and nature of life that does not require a creative deity. What is truly remarkable and far more disturbing is that the anti-science and anti-rationality banners have been raised by reputedly progressive Gaian and New Age proponents, in whose ranks appear distinguished scientists and intellectuals.

Environmental activists by the very character of their work must rely, unless they themselves are trained scientists, on the findings and opinions of scientists. But often, like government regulators themselves, they tend to pick and choose those opinions which most closely reflect their own preconceptions and beliefs so as to bolster their personal predilections and of course avoid the possibility that their opinions will not hold up under scientific scrutiny. It is unlikely that admirers of Teilhard de Chardin and his fanciful "noösphere" concept would read anything by Ernst Mayr, the 20th century's most brilliant evolutionary biologist, or for that matter the late Stephen Jay Gould, the political Left's smartest booster of evolution.

What is the reason for this? Why did generations past, and now individuals of our time, feel the need to attribute existence to a creator or a plan? Was it just the preachings of organized religion, or something more fundamental, something so elementary that it was shared by animist religions, and today by post-Enlightenment materialists and scientists who may not even believe in a god?

Certain words and concepts re-appear in the scientific and philosophical literature about evolution and the origins of life, the most frequent being the word randomness. Apparently it is extremely difficult for even secular scientists to accept the notion that we humans, or life on earth in general, are a product of chance, and that, had the dice turned up differently, we might not be here at all. Humans can and do live with tragedy and misfortune but they seem to have profound difficulty living with uncertainty, which is implied by randomness. It then becomes more comfortable and safer to posit the idea of a plan and necessity.

For practical as well as ecological reasons, whether the universe was born out of pure chance is irrelevant. We don't have to know the "Why?" but rather the "How?"; we need simply to understand that once the primal elements and conditions were in place, life appeared, evolved and persisted because of specific biological and physiological mechanisms and processes, including replicating DNA. It is a certainty that we all possess DNA and that our descendants will possess it and their descendants as well, and that evolution of life forms will continue (that is, if we do not commit ecocide) with a certainty based on these fundamentals. We need to know *what* we must do to insure the perpetuation of these conditions. Thus, our social activism needs to be grounded in both science and ethics, in an understanding of the conditions under which evolution functions, and in adoption of social policies that will preserve these functions.

Ω

Ecology and Evolution

Motivated by the desire for progress, humankind has sought to shake loose its biological constraints and evolutionary history. From the advent of agriculture through the industrial revolution into the space age, and now in the age of computers, humans have sought to exploit Nature through the control of land, resources, institutions and other humans. The undesirable side effects have been disregarded, minimized or rationalized as inevitable or necessary: illness and death from nuclear, chemical and military technology, profound alienation from Nature, disintegration of the planetary systems upon which human survival depends, and subjugation of the masses of people needed to support the gigantic industrial enterprise.

So it is not surprising that a reaction to this suicidal way of life set in: the environmental movement, determined to mitigate or reverse the effects of industrialism and its global culture. The movement's origins go back to the 19th century as does the concept upon which it was based (Ernest Haeckel coined the word 'ecology' in 1866). Since that time, its meaning has diversified, connoting, at different times and to different people, a field of academic study, natural resource analysis, a foundation for the environmental movement, spiritual inspiration for New Age, bioregional and other social change groups, a red flag to corporations and developers, and, most recently, a rallying point for the Green political movement, alienated as it is from traditional politics and seeking a new societal paradigm.

The word *ecology* derives, as does the word economics, from the Greek *oeikos*, meaning habitation or house, and signifies the study of the interrelationships of living things with each other and their physical environment. The concept of ecology arose out of the fact of evolution,

24

which itself arose not from any one discovery or singular insight but from the intertwining strands of several scientific disciplines, notably comparative anatomy, embryology, geology and paleontology. What gave impetus to the concept of evolution were in fact the lively currents of 18th and early 19th century theories and discoveries about the age of planet Earth, the fossil record, growing suspicions about species extinctions and appearances, and the intellectual contributions of many people from Darwin's time or just before it, in particular Charles Lyell, Jean Baptiste Lamarck, James Hutton and Georges-Louis Leclerc Buffon as well as Darwin's grandfather Erasmus Darwin.

Neither evolution nor its postulated mechanism, natural selection, was proposed out of whole new cloth by Charles Darwin; Darwin pulled together fossil evidence, his own experiments with artificial breeding, geological substantiation, and variation in nature, showing through his brilliant synthesis that existing life forms arose from earlier simpler forms, and that their complexity and novelty could be explained without the need for design or creative force.

Were ecology limited solely to the biological and physical relationships of living things, it would have limited relevance to human social concerns. But because human culture—its origins and development, religious beliefs, ethical choices, technologies, social behavior, mores (all originating from the human brain and intellect)—is a product of both biological and social transmission, and because all life forms are inextricably embedded in Nature, ecology is charged with trying to understand how human society is an outgrowth of Nature, how in turn it affects the rest of the natural world, and in particular how it attempts to escape biological constraints.

Ecology takes its meaning and substance directly from the fact of evolution. The concepts of ecology and evolution emerged almost simultaneously; Haeckel's term followed the publication of Darwin's *The Origin of Species* by seven years, a book written before Gregor Mendel's genetic experiments, conducted in a Moravian monastery, were discovered in the early 20th century. They are really two sides of the same coin.

Evolution deals with the origins and development of species; ecology is the study of how species live. Evolution studies life forms over time; ecology

25

studies them in space, in specific locales. Evolution occurs through the interaction of genes and environment; ecology is the study of those interactions.

An understanding of how organisms and species interact, how biotic communities and ecosystems function and what limits and requirements are placed on living things is not possible without an understanding of evolution and its postulated but accepted mechanism, natural selection. The biological, chemical and physical linkages between and limits imposed upon organisms and species are expressions of the now-accepted truth that any given species (actually, its component populations) is a product of the interaction between its genetic inheritance (genome) and its environment.

Living creatures look and behave the way they do because of their genomic plan and because their physiological and behavioral traits—their lifestyles as it were—are well suited to prevailing external conditions. The requirements, constraints, enemies and opportunities of a species and how it acts to achieve, cooperate with, take advantage of or avoid these are, in the truest sense of the word, the 'ecology' of that species. Organisms behave as they do because of evolutionarily tested inherited tools and learned behavior, or, in the case of humans, with chosen or habitual values, culture and education, often superimposed on (or in conflict with) inherited tendencies.

None of these acts are performed in a vacuum but rather in a concerted, though not conscious, way that affects and is affected by other members of an ecological consortium, something explicitly noted by Darwin in *The Origin of Species*. In many fascinating cases, we see examples of mutualism, symbiosis or co-evolution, in which certain crucial life functions of two species are mutually fulfilled; close study usually reveals a benefit to both organisms that makes such behavior conducive to the reproduction or survival of the individual (or population) involved. Co-evolution is an expression of extreme interdependence that unites species irrevocably and indefinitely, and in most cases is indispensable to the survival of both species.

But co-evolution or mutualism can be more general. An individual or species can alter its physical environment incidentally, thereby creating new environmental conditions that are favorable to the continuation of that species, which continues to flourish in that habitat, its presence and behavior insuring continuation of the new environmental condition. But the reverse can happen too; a seemingly insignificant change in the environment can drastically alter conditions for a highly specialized species and lead to its extinction.

Such behavior occurs because individuals (actually a group of interbreeding individuals within a species called a population) of a particular species possess a set of genes, a genome, that is adaptive, meaning that in the present environment it is conducive to the survival and reproduction of those individuals. Thus, the adaptive genome is rewarded with reproductive success and is perpetuated in future generations, providing it remains appropriate to external conditions. (The genome influences both physical traits, called the phenotype, as well as behavior).

Of course external conditions do change over time due to natural events and the impact of living things on the biological community or ecosystem. Natural selection, acting on individuals, maintains the favored genome while the environment is stable, or in the event of environmental change, favors those individuals (or population) with slightly different traits that are more adapted to the new conditions.

Natural selection simply means that those individuals possessing the more adaptive genome traits reproduce more successfully, leaving more offspring, and thus spread those traits through the population. The amazing power of even a small competitive edge in reproductive success was noted by archaeologist Ezra Dubrow in *The Human Evolution* (ed. Paul Mellars and Chris Stringer). In his study he found that with only a slight decrease in life expectancy in a Neanderthal population of 1% to 2% and a comparable increase in a competing *sapiens* population, the Neanderthal population would become extinct in only thirty generations or less than 1,000 years. Natural selection in stable environments tends to conserve adaptive genomes rather than opt for untested ones. It is not the popular "nature red in tooth and claw" of pop evolutionary theory, not a battle but rather a sorting out of more adapted from less adapted individuals. Scientists simply call it differential reproductive success.

New life forms and their populations have, seemingly, evolved so as to utilize energy and resources more efficiently, without waste or conflict. The result is growing biodiversity, an increasing diversification of forms suited to particular niches, each one occupied in its entirety by one form only. A niche is not merely a geographic location but a broader abstract area that includes that species' predators, prey and habitat requirements. Large general food sources such as seeds and insects can of course be shared but often they are divided up in ways not immediately discernible to us. A variety of shorebird species can feed together at the water's edge on quite different things—some on small mollusks, some on crustacea, some on insects, some on small worms and fishes, and at different levels: on the surface, on the bottom, or deep in the mud. Many tropical birds inhabit different levels of the forest; some birds are specialized to eat certain seeds—the crossbill for example, has a peculiar twisted beak adapted to removing conifer seeds. Climate changes, natural disasters, drought, vulcanism, disease, and man-made changes, as well as the activities of living things themselves, sometimes open up new niches which then become available for colonization. And populations within species that have particular attributes or adaptations may split off from the rest of their species to exploit a new habitat and eventually become a separate species as their behavior and traits diverge.

Evolution Misunderstood

Gross misunderstanding of the fact and mechanisms of evolution continues today and Darwin remains a favorite scapegoat for all manner of political ideologies. Jeremy Rifkin, critic (and justly so) of genetic engineering and modern technocracy in general, in his *Declaration of a Heretic*, says: "According to the law *(sic)* of natural selection, the story of evolution is a story of rank opportunism and utilitarian self-interest. These are the overriding principles that govern the very fabric of life . . . For Darwin, maximizing control over the environment becomes the *sine qua non* in nature as in society".

Sad to say, Rifkin has seriously misunderstood evolution. First, there is no "law" of natural selection; it is simply the postulated mechanism by which evolution proceeds. Second, though adaptive genes/genomes (and in some cases fortuitously pre-adapted ones appearing in organisms prior to environmental conditions to which they are suited) may infer "opportunism", characterizing them as "rank" imposes a human value judgment on value-neutral processes. The prevalence of certain genomes is neither "rank" nor timid; it merely signifies the non-prevalence of unsuccessful ones.

Third, the concept of "utilitarian self-interest" suggests individual consciousness on the part of animals and plants. Animals and plants do not murmur to themselves that they are the most fit and set out to reproduce themselves with a vengeance. They behave as a result of a genetic program whose adaptive value has already been successfully demonstrated by the fact that the extant individuals exist and reproduce successfully. It is only through this demonstrated adaptiveness of the genome that the "self-interest" of the individual is served. Thus, the overriding principle of evolution that governs "the very fabric of life" is not self-interest but adaptation.

Rifkin also ignores the environment's role in natural selection and the fact that changes in organisms or populations are "good" or "bad" only in relation to existing environmental conditions. It may be a matter of luck if certain genes are adaptive but it is a demonstration of adaptation if they persist and spread.

Darwin's simple thought, that natural selection improves individuals only in relation to others in the vicinity ("As natural selection acts by competition, it adapts and improves the inhabitants of each country only in relation to their co-inhabitants."), contains the essence not only of Darwin's theory of evolution, the force resulting from the interplay of biota and the physical world, but the essence of ecological thought by offering the perspective that not only do living things interact with each other and the external world but that a change in one component will affect others.

Rifkin's statement that, for Darwin, "maximizing control over the environment becomes the *sine qua non* in nature as in society", is totally inapplicable to evolution. Organisms interact with and often change the environment but only in the context of natural selection, which "rewards" adaptation. Thus, it is adaptation that is the *sine qua non* of evolution, permitting the spread of suitable genomes through increased reproductive success. To equate the exigencies of natural selection—adaptation and conserving of the status quo—with human control over the environment is to mistakenly confuse biological evolution, guided by genetic programming, with cultural evolution, which is governed by human choices and social transmission.

Cultural evolution is a matter of conscious choice and therefore based on reigning societal values; in the case of modern industrial society, it just so happens that these widely transmitted values run counter to the exigency of adaptation, a destructive luxury that non-humans can ill afford. In non-humans, maladaptive behavior usually means death of the individual or disappearance of the population or sometimes extinction. In the case of human cultural choices, however, non-adaptive choices are frequently (but only temporarily) insulated by such things as medical intervention, social welfare, government fiat, scientific rationalization or mitigating technology.

Certainly Darwin, living in the early industrial era where radical social stratification was the norm, was as much a product of his time as nuclear physicists are today, and his reading of Thomas Malthus' theories on population and resources played a crucial role in his understanding of how organisms and species in nature partitioned out or competed for resources and living space. But the influence of Malthus on Darwin was only one of many influences; even the notion of organisms competing for resources pre-dates Malthus, who put such ideas in a more formal mathematical context.

One of the fundamental misunderstandings of our day is the erroneous interpretation of natural selection, which appears to allow a particular organism or species to benefit at the expense of another. But it is really more accurate to say that a *potential* organism "benefits", if that is the proper word, at the expense of another potential one, in the sense that some individuals are reproductively more successful than others. Even if this were not so, however, the notion that all potential individuals—that is, all the germ cells of all living things—are entitled to survive is absurd on the face of it, unless one denies the obvious fact of and need for the death of individuals and much of their genetic potential. Not all germ cells germinate.

What was most deplorable in Darwin's time and after was the embracing, by advocates of social engineering, of a highly distorted version of Darwin's theory of natural selection (a theory simultaneously proposed by Alfred Russel Wallace), in order to shore up prevalent *a priori* socio-political theories, notably the attribution of inferiority to nonwhite races, defective individuals and those of low economic status. Darwin's brilliant thesis of natural selection was subverted into the proverbial "nature red in tooth and claw" and extended to an industrial, highly stratified, class-conscious society to justify the social and economic status quo. Evolutionary biologist Ernst Mayr suggested changing "Social Darwinism" to "Social Spencerism" to indicate the influence of Herbert Spencer and his supporters.

The excesses of the Social Spencerists may have unwittingly influenced social science studies of our day too. Evolutionary studies have been conspicuously absent from sociology and social studies dealing with human behavior and cultural practices. The discipline of sociology rarely if ever looks at the influence of heredity or genes and still in many cases regards human behavior and relations as almost totally socially determined; the genetic influence that, they freely concede, guides non-human species is not acknowledged in humans. It is arguable that sociobiology (now called "evolutionary psychology") is what we got when the social sciences ignored evolution.

It is likely that this attitude towards the study of the human species originally contained an implicit assumption that humans were independent of the laws that govern the rest of the natural world, a view that leads easily towards not only alienation but an instrumental view of nature and other life forms. As such it could be argued that this attitude is the precursor of the objectification and commodification of nature and a major cause of our environmental crisis. Non-human species by this measure can be considered "legitimate" objects of scientific study, subject to certain "laws" while humans remain apart from and above these laws. (One wonders at what point in history we became "fully" human, free from the biological constraints that were presumably operative for our primate ancestors).

More recently there have been rumblings of discontent from surprising corners and schools of thought made uncomfortable at the notion held by some conservatives that natural selection in Nature is the biological equivalent of economic competition in the so-called "free market". A superficial comparison might well mislead some to perceive a parallel, but a closer look at the "free market" concept lays the notion to rest.

A market is not a single organism, for one thing. It is a collection of disparate parts, some related, some not, consisting of artificially manipulated relationships. Second, the market has no genetic endowment and is purely a human invention and concept. As such it is a cultural artifact, affected by unpredictable factors like advertising, rumors, public relations, consultants, as well as by government policy and Wall St. Third and most important, there is in any case no real "free market" since conditions like monopoly, collusion, subsidy, corruption due to corporate-induced shortages (such as those induced by California energy suppliers in 2001 in order to raise demand and consumer prices), and in general specific actions of individuals within the market, all demonstrate that supply and demand are not at all freely determined and that manipulation of consumers by individuals and institutions within the market conspires to defeat any notion of a free market.

While it is true that the pop media love to grab at straws of a good controversy to fill their pages, the truth is that there is no scientific basis—either in the opinions and work of those on the Right or on the Left—for any such thing as "genetic determinism" and those who dedicate themselves to rebutting such determinism have misplaced their discomfort and political rhetoric. Why, then, such hot denials? It seems likely that they are stimulated by the incredible advances in genetics with regard to the human genome, coupled with the undeniable fact that pharmaceutical companies and agribusiness are intent on monopolizing genetic (primarily seed and plant) resources by manipulating genes in order to create "novel organisms" that can then be patented.

This latter concern is a genuine one and needs to be counteracted in the social, political and economic spheres. But it should not be done with a broad brush that inadvertently tarnishes honest science. Distortion, hyperbole and the raising of hypothetical risks do a disservice to both science and genuinely progressive social movements opposing the proliferation of genetically modified plants and animals. Evolutionary principles are on the side of such causes and can only help them rather than hurt them.

Even with the human brain, culture, morality and technology, a premium is still placed on adaptive behavior just as evolution does with all living creatures. In industrial societies, brute strength, size and phenotype are relatively unimportant in the human species. But if humans do not adapt their behavior appropriately and consciously—that is, recognize what behavior and environmental conditions threaten their existence—there is no reason to believe that humans will not eventually pay the same price as all non-adapted species: extinction.

The competition that does occur in Nature can occur between individuals or between species, which may lead to the recession of certain traits, extinction of a population, or its evolution over time into an entirely new species if it becomes geographically isolated. The lesson of adaptation in modern humans is not a lesson about social, economic or racial dominance but a lesson about ethical, cultural and technological choices: not natural selection but social selection, and as such, as relevant to our survival as whether certain field mice adapted, reproduced and passed on their genes instead of others.

The core of evolution's lesson is adaptation. To meet changing external conditions (climate, food sources, enemies, shelter), living creatures must modify their behavior or become physically adapted. The essentially infinite diversity (number of possible genetic combinations) in animals and plants, plus the imperfections that cause genes to mutate randomly or be improperly duplicated, means that within any given population there will be individuals whose genetic makeup is particularly well suited to the existing environment. Such individuals will, under these conditions, reproduce more successfully and pass on their genome, until they are spread out in the population—natural selection at work, differential reproductive success, not violence or bloodshed. The results of natural selection are manifested not here and now but in the future—by offspring. Moreover, evolution and natural selection cannot predict the future and so natural selection acts only in terms of present conditions and presently available genes. (This is the basic fallacy in James Lovelock's Gaia hypothesis, which argues that the earth is a living organism that regulates and maintains the conditions necessary for life; such an argument obviously fails because present forms of life are suited only to present environmental conditions).

Ω

Evolution's Balancing Act

Evolution is the appearance of new life forms from earlier ones, what Darwin called "descent with modification". It is a constant tug-of-war between preserving existing life forms and eliminating the less fit, between conservation and diversification, between conservatism and opportunism. The balance is reached through natural selection whereby better adapted individuals out-reproduce the less adapted. The word "selection" is misleading because it suggests a selective force; a more accurate description might be "natural elimination". Actually neither word is precise because it is not individuals *per se* that are being eliminated but their reproductive potential. The process is called differential reproductive success.

Existing adaptive traits, which are expressions of inherited gene combinations, are favored over randomly appearing untested new ones. Unexpressed genes in the genome may mutate, recombine or simply lie latent until new environmental conditions place new demands on the population. But as long as conditions are stable, the trend for species, or rather populations (reproductively unified groups within species) is to remain the same.

Evolution sorts out the possibilities offered by genomes to achieve workability with variability: never perfect, never predictable, never fixed, hence the tension between stability and flux. Nor does Nature concern herself with the death of an individual; that is left to the human species, though seldom accompanied by an equal concern for extinction of non-human species.

Life is neither as fragile or unresilient as is often believed. Given an external energy source, (the sun), one could almost say that life (metabolizing, self-regulating, self-replicating organisms) was inevitable. When one looks at harsh environments such as polar icefields and parched deserts, and at the aftermath of natural disasters such as volcanic eruptions, floods and fires, one is immediately struck by the persistence and reappearance of life. As repeated resurgences of plant disease, blights and insect predators demonstrate, even wholesale insecticide spraying has little effect except to kill off weaker insects and allow resistant ones to take over.

Only a radical change in the basic conditions under which life originally evolved can eradicate life: synthetic chemicals (which destroy fundamental hormone and enzyme systems in both human and non-human species), ionizing radiation which, like many synthetic chemicals, introduces random genetic changes in our genes that are mainly deleterious, or removal of all sources of air, water and energy (death of the sun). Most important, the major environmental threats we face today—radiation, synthetic chemicals, and loss of biodiversity—threaten the ultimate survival of our species rather than non-human ones.

Over the course of evolution the rate of appearance of new species exceeded the extinction rate of existing ones but now this process has been reversed due to the accelerated destruction of habitat, the most definitive means of exterminating species. This is readily understandable in ecological and evolutionary terms, since living creatures are adapted and tied to their physical surroundings and have no real identity apart from their habitat. One could almost define a species by saying it is the physical result of the interaction of a genotype, through its phenotype, with its environment. Some species never had to face changes in their environment; some we consider "primitive" such as the horseshoe crab have remained constant for hundreds of millennia. But all the while geological, climatological and physical changes have continued to open up new niches, enabling new species to evolve to fill them.

Said anthropologist/poet Loren Eiseley:

> "Form, once arisen, clings to its identity. Each species and
> each individual holds tenaciously to its present nature. Each
> strives to contain the creative and abolishing maelstrom that
> pours unseen through the generations. The past vanishes,

the present momentarily persists, the future is potential only. In this specious present of the real, life struggles to maintain every manifestation, every individuality, that exists. In the end, life always fails, but the amorphous hurrying stream is held and diverted into new organic vessels in which form persists, though the form may not be that of yesterday". (Loren Eiseley, *The Star Thrower*, 1978).

However, random errors in gene duplication and mutation of single genes can occur to enhance the latent genetic diversity needed for populations to draw upon in the event of environmental change, i.e. needed for the process of natural selection to operate. This genetic diversity is crucial to evolution because it provides a broader range of potentially adaptive gene combinations and traits. Were Nature entirely perfect in its workings, all members of a species would be uniform, genomes would never change and the species would be a collection of static clones, at extremely high risk of extinction in the event of adverse environmental change.

Ω

Sidestepping Evolution

Because human survival has been narrowed down to an issue of conscious and cultural choices, the rapid destruction by colonialism, industrial expansion, resource extraction, land development and agribusiness, of native tribes, communities and cultures—what John Papworth calls the Fourth World of self-sufficient, subsistence human communities within nation-states—is of global ecological consequence.

These cultures are the human equivalent of animal and plant populations because of their geographic, cultural and reproductive isolation. As relatively isolated populations adapted to and enmeshed in their physical environment, they are precious, irreproducible reservoirs of both genetic and cultural diversity, that same diversity which Nature goes to such lengths to preserve in order to provide all species with the ability to adapt to new environments.

Because all species are products of gene/environment interaction, species that become extinct can never re-appear because the exact environmental conditions out of which they arose long ago can never re-occur. Thus, populations, human or otherwise, must be preserved in their habitat, with large enough numbers of individuals, to allow the perpetuation of demonstrably adaptive genomes (as well as different untested ones for the uncertain future). Thus, the protection of native peoples, tribes, cultures and societies in their natural habitat—in other words, the preservation of maximum biological and cultural diversity—is both ethically and ecologically necessary.

In *The Last First Contacts*, Jared Diamond notes that whereas modern-day Europe has fifty languages, most in the Indo-European family, New Guinea, with under one-tenth the area of Europe and less than one-one hundredth its population, has about a thousand languages, many unrelated to those spoken elsewhere. Regarding this loss of diversity, Diamond says:

> "We wouldn't mourn the shrinking cultural diversity of the modern world if it only meant the end of self-mutilation and child suicide. But the societies whose cultural practices have now become dominant were selected just for economic and military success. Those qualities aren't necessarily the ones that foster happiness or promote long-term human survival. Our consumerism and our environmental exploitation serve us well at present but bode ill for the future. Features of American society that already rate as disasters in anyone's book include our treatment of the aged, adolescent turmoil, abuse of psychotropic chemicals and gross inequality. For each of these problem areas, there are (or were before first contact) many New Guinea societies that found far better solutions to the same issue . . . Unfortunately, alternative models of human society are rapidly disappearing and the time is almost past when humans could try out new models in isolation". (*Natural History*, August 1988).

What are the threats to biological and cultural diversity? They are worldwide and spreading rapidly: in addition to the militarism and economics mentioned by Diamond above, they are: uniformity (food monocrop culture), behavior conformity (needed to maintain the power and legitimacy of entrenched elites), mass culture (needed to spread common-denominator non-threatening values, stimulate consumption and discourage heresy), technology (to eliminate self-sufficiency and self-determination and create dependency), genetic manipulation (to insure homogeneity, pre-empt demands for environmental cleanup, allow corporate control of basic genetic resources for food and medicine, and facilitate social engineering experiments by the scientific elite), and the dominant role of the central state (to mandate and enforce all of the above).

Because we can shape the physical world with technology, we somehow think we have sidestepped the basic laws of evolution and ecology. But it is precisely because all species are the result of the interaction of genes and environment that our failure to cultivate adaptive behavior and choices may lead to our extinction. Unlike non-humans whose behavior is genetically regulated and influenced by natural selection, we make conscious choices of behavior based on habit, dogma, religion, politics, superstition, greed, or spite—or hopefully intelligence—while we ignore signals that would cause aversion behavior in other creatures (foul water, tainted food, chemicals).

Modern technology has imposed human needs on the rest of the natural world, thereby undermining or replacing the self-regulating mechanisms that arose through natural selection. Natural systems, characterized by self-regulation, stability and sustainability, are being sacrificed and replaced by arbitrary, imposed conditions in the interest of human profit, convenience or amusement, rather than in the interest of the organism itself.

In agriculture, for example, we manipulate plants that in the wild evolved under certain natural selection pressures and evolved their own set of adaptive traits, but instead of concerning ourselves with the plants' needs, we impose criteria during breeding in order to meet human demands such as rapid growth, size, appearance, shape, transportability and productivity, traits that would not necessarily have evolved in the context of their total needs. Thus we sidestep evolution and natural selection and scratch our heads when new undesirable traits or disastrous side effects appear.

In addition to the environmental and health effects of the sidestepping of evolution, there are, additionally, social and cultural effects. The replacement of self-regulating systems that evolved via natural selection with artificial, human-imposed ones leads to technical and institutional problems that in turn require external regulating mechanisms. These may be hazardous, inappropriate, inadequate, fallible or malicious. Where human short-term benefit or profit has priority, artificial regulation becomes imperative: fertilizers, pesticides (to counteract the effects of counter-evolutionary agricultural practices), bureaucrats, lawyers, scientific analysts, interpreters, mediators, managers, pollution control experts. Thus, the management of biological impacts must be extended to create a socio-political system of management that permeates nearly every aspect of our lives.

Ω

Biodiversity

The global biodiversity crisis could well be considered the gravest crisis we face on earth, in that it undermines both the processes and products of evolution. Paul Ehrlich has compared the loss of biodiversity to the removal of rivets on an airplane. The loss of biodiversity at all levels—genetic, populations, species, ecosystems—plus the human modification of plant and animal DNA and the DNA-damaging release of synthetic chemicals and ionizing radiation may not cause an immediate crash of Spaceship Earth but cannot be regarded as anything but lethal. In any case it is the most profoundly counter-evolutionary trend in the history of humanity. If blasphemy is defined as an irreverent defaming of ideas or objects held sacred, then the extinction of species and the rearrangement of DNA, the material requirement and precursor of evolution, most assuredly fit this description.

Biological diversity encompasses genetic diversity, population diversity, species diversity and ecosystem diversity, and is both the precondition and result of evolution. The near-infinite genetic combinations available in living things serve various functions, not least of which is providing sufficient variety for the hedging of bets when environmental changes occur. In turn, adaptation to environmental change facilitates speciation; a population within a species that inhabits a different geographic area may, in response to environmental pressure, eventually become a separate species.

Diversity of species is not only a result of evolution but a way to partition out resources and functions within a biotic community or ecosystem more efficiently and less stressfully. Species diversity is an indicator of planetary health in that the more species there are, the more different habitats we know exist. Species extinctions indicate that the planet's ecosystems are being destroyed, along with the impairment or loss of their functions, and the earth is becoming more homogeneous. Ecosystem diversity can provide redundancy for species, be a source of new individuals when a local population declines and possibly offer refuges or alternatives when some systems fail.

While extinction is the fate of all species (99% of those that ever lived are gone), this statement is really saying something positive: evolution. And evolution is really just another way of saying speciation. Evolution, speciation and biological diversity are all about the same thing: the disappearance of the old and the appearance of the new. If Nature has her way, more new species evolve than go extinct. In the past 100 million years, notwithstanding cataclysmic extinction events, biodiversity continued to increase because of the richness of tropical rainforests and the fragmentation or division of land masses into smaller ones that offered new niches.

Traditional ecological wisdom holds that species diversity is a stabilizing factor in ecosystems and makes them more resilient and able to recover from environmental adversity, but it may not be true under all conditions. It is probably true for intact ecosystems free of human interference. In undisturbed tropical forests, old age, disease and weather constantly uproot tall trees which then fall and take other trees and plants with them, opening up the ground to sunlight, which then enables some plants to gain a foothold formerly denied them by the absence of sunlight. But enough previously cleared areas have grown up elsewhere to shade out opportunistic invaders so they never gain overall.

An important concept is that of keystone species which, if removed, will lead to drastic changes in the rest of its biotic community. These other species may decline or go extinct, or become super-abundant, or species formerly excluded are now able to enter. Edward O. Wilson says the sea otter may be the most formidable keystone species and refers to the overhunting that nearly extinguished it in the 19th century. This caused an explosion of the otter's chief prey, the sea urchin, whose population then exploded, to feed on ocean kelp to the point where the kelp forests under the sea rapidly declined. A restoration effort in the 20th century has helped to restore the otter and ultimately the original ecosystem.

But human-induced change by bulldozers and contaminants is on a larger scale and produces changes in many more places over a wider area and at an accelerated rate, with regional and, indeed, global implications. While evolution can repair local ecological upheavals (some scientists postulate that some habitats flourish under a regime of localized disturbance), large-scale human-induced change can easily outpace evolution, creating artificial conditions for so long a period and over such a wide area that indigenous plant and animal populations disappear, ecosystems disintegrate, and secondary and tertiary effects ensue.

Genetic diversity can only be maintained through sufficiently large numbers of individuals and populations within species, which in turn require large enough areas of appropriate habitat. Since all populations suffer mortality, there needs to be a source of new influxes to maintain the numbers of individuals and their pooled genetic diversity. If the number of individuals within a species (or population) falls below a critical threshold, it can be threatened by hereditary weaknesses and disease, which are magnified by inbreeding. It is believed that this is what has endangered African cheetahs, which were once part of a family of three species that were spread across Asia. The key to all this is preserving species by preserving their component populations in their habitats, not in zoos or small preserves but in large contiguous tracts of undegraded land.

Why preserve biodiversity? Numerous reasons can be cited: aesthetic, educational, medical, scientific, economic, ethical. But there may be two overriding reasons. Ecosystems are in effect the earth's life support systems. They provide basic services indispensable to all life on earth: climate regulation, nutrient recycling, decomposition of dead organic matter, resources for food, shelter and medicine, minerals for industry and technology, pure fresh water, control of pests and blights, and pollination of flowering plants and food crops.

The other reason is that more likely than not we have an innate need for wildness and diverse wild landscapes, with which our species evolved. This ancestral biotic diversity was our natural environment and therefore a stimulus to our evolutionary as well as cultural development. Human evolution has taken place over several million years, during most of which time (all except modern history in fact) humans lived in small hunting and gathering groups. Our evolutionary history took place in a wild, green natural landscape of tremendous plant and animal diversity. It is therefore highly likely that the human need for and appreciation of nature and wilderness is innate. Too little time has elapsed since our hunter-gatherer days to have brought about evolutionary changes that would have eliminated our instinctive need for nature. There was no way and no reason for natural selection to act on our predilection for wild nature. Workplace fatigue and boredom would therefore seem to be not purely psychological responses but biological responses to being deprived of natural surroundings.

In fact, our superb ability to adapt culturally to adverse conditions conceals to a large extent these innate, evolutionarily derived needs. Disneyworld surrogates may fool small children but not our brain or psyche. Unfortunately the human constructs of material well-being, affluence, entertainment and recreation, while distracting compensations, are not satisfactory substitutes for the loss of natural landscapes, ecosystems and life forms with which we evolved. Those who believe space colonization is feasible need to give serious thought to whether such efforts are inherently doomed to failure on evolutionary grounds. Modern technology has not freed us from our dependence upon the free earth services or our innate memory of unspoiled nature. To continue lopping off branches of the evolutionary tree is not only a defilement of the greatest cosmic achievement, evolution, but could result in lopping off the branch we humans sit upon. Once lopped it will not re-grow. Treasuring biological diversity in its smallest or least appetizing details may be the only means we have of preserving not just our lives but our own humanity.

Ω

The Threat of Genetic Manipulation

Many ideologues have sought to use Nature as an example for human behavior and social structure, though often the "Nature" they refer to is a seriously incomplete or distorted view that, unsurprisingly, supports their biases perfectly. Eugenics, for example, a socially regressive school of thought that flourished into the 1930s, represented the artificial (that is, social as opposed to biological) equivalent of natural selection, or so it was regarded, and not just by people like Hitler with his theories of racial purity. Not all the eugenicists were racists or Nazis; some believed that the sterilization of what they termed the "unfit"—those with physical or mental defects—would improve the human species.

The genetic manipulation endeavor represents the next phase in a political process that turns democracy on its head and represents the biological equivalent of the nuclear energy experiment, namely introducing random changes into an orderly integrated entity, the human genome. Powerful prevailing interests apparently—first nuclear physicists and now molecular biologists and their corporate sponsors—need to characterize their work as both benign and necessary.

The by-products and "externalities", both social and ecological, are downplayed or declared amenable to mitigation, and the uncertainties and unknowns are not mentioned, while the doubts and fears of the public (essentially everyone who will not stand to benefit directly) are dismissed as the work of fearmongers. Public debate is stifled to the point where the real questions never get asked, much less answered. In the case of ecology and evolution, the large questions that need to be asked are: what kinds of technologies are both ethically *and* evolutionarily appropriate? Which

46

technologies will fulfill genuine human needs while preserving biodiversity and allowing the processes of evolution to proceed without interference?

Today we have a new version of eugenics in the form of genetic engineering, wherein scientific elites and corporations, using the argument of feeding the starving masses and curing hereditary disease, seek to modify the life code, DNA, to achieve their supposedly noble objectives. There are both biological and ethical reasons to resist. Such efforts not only sidestep evolutionary processes but will probably fail because the premise of genetic modification is based on a radical reductionism: the belief that one genetic change will produce the desired effect, accompanied by the faith that enough is known about genome functioning and the proteins they code for to predict the outcome, a belief that such intervention is inherently harmless, and a confidence that we have sufficient knowledge of how evolution works to make predictions.

In fact the body of evolutionary knowledge as well as recent experiments indicate the exact opposite. The knowledge that organisms and species evolved from pre-existing ones with a long evolutionary history, the fact that species are a result of the interaction of genes and environment, and the fact that a genome is more than a collection of individual genes but represents an integrated entity in which one small change can create a large result, should give us warning that most of the claims made by the pharmaceutical and chemical corporations are nothing more than speculation and wishful thinking.

We are on the verge of a genetic revolution that proposes to literally reconstruct genomes, that is, to recombine DNA to serve human purposes. But substituting intentional alteration of genetic material for the force of natural selection is not only dangerous and fraught with uncertainties but is truly blasphemous in evolutionary terms. Our species has not yet demonstrated that it possesses the moral judgment to conduct what Nicholas Wade terms the "ultimate experiment"; the risks of such experiments are unknown and probably unknowable; the usurping by humans of the powers of evolution can lead to either complete societal control by the scientific and bureaucratic elites or to bumbling ineffective totalitarianism.

There are of course sound scientific reasons for avoiding genetic alteration. Random changes in genes are seldom beneficial, first because they occur outside the context of natural selection, and further, because the genetic makeup (and hence the organs and systems) of organisms evolved together in a harmonious whole, and finally because, as stated before, such alteration substitutes human needs for adaptive tested ones. Even if only one change is made, it may have repercussions in an entirely different organ or system because the organism and genome evolved as a whole, not as a collection of disparate parts.

For example, much is made of genetically altering certain plants to give them nitrogen fixation capability such as leguminous plants possess, using bacteria co-evolved with plant rhizomes. Recent experiments attempted to insert a Brazil nut gene into soybeans to enhance its nutritional properties by adding missing amino acids, but it was then discovered that those who were allergic to Brazil nuts were having allergic reactions to the genetically modified plant (J. Nordlee et al; *New England Journal of Medicine*, 1996). Since the metabolic systems of non-leguminous plants evolved as a balanced whole to take up nutrients without nitrogen fixation, the results of inserting a gene (and rarely is a trait caused by a single gene) for nitrogen fixation cannot be foreseen. If we decide to substitute human-imposed needs for the tedious force of blind natural selection, it becomes even more crucial to make choices conducive to the health, reproductive capacity and survival of the species in question not to mention human health.

More recently, an experiment in natural breeding, not transgenic, to produce a strain of docile fox by selecting and breeding the tamest ones of a group brought about a number of seemingly unrelated physiological changes in appearance, reproductive cycles, timing of physiological development, and decline in adrenal gland production of hormones. Dr. Darcy Morey, anthropologist at the University of Kansas commented, in the *New York Times*, 3/30/99: "Clearly, the link between modifying the neurology of an animal and its biochemical balance is all intertwined with other aspects of physiology."

The director of the experiment, Dr. Lyudmila Trut of the Institute of Cytology and Genetics at the Russian Academy of Sciences also stated: "Behavioral responses are regulated by a fine balance between neurotransmitters and hormones at the level of the whole organism. Even slight alterations in those regulatory genes can give rise to a wide network of changes in the developmental processes they govern." She stressed that changing a complex of genes even by selective breeding is not necessarily beneficial and that even slight alterations "might upset the genetic balance in some animals, causing them to show unusual new traits, most of them harmful to the fox." If these adverse effects can be caused by mere traditional breeding within species, then clearly transgenics involving moving genes from one species to another raises the same issues and risks, and even more so.

If society ultimately acquiesces to genetic engineering in order to substitute human-imposed demands for the tedious force of blind natural selection, it would seem to be self-evident and crucial to make choices conducive to the health, reproductive capacity and survival of the species in question, because we are preventing natural selection from making an adaptive choice. But the as-yet-unasked and unanswered question is whether we will ever have enough information to make such choices. Many of today's debates over technologies such as nuclear power and genetic engineering revolve around the adequacy or accuracy of information and how much uncertainty society is willing to accept.

One spreading problem, that of invasive alien species which can opportunistically invade disturbed areas and displace native species, has not been given sufficient attention in the argument about transgenics. Under normal circumstances, alien species cannot compete with native species, but ecological disturbances often alter habitats severely and create a niche where alien species can move in. Two such plants, Water Hyacinth and Purple Loosestrife, are seriously threatening freshwater wetlands in the US, and, being alien species, have no natural predators or controls.

Transgenic manipulation of species—the insertion of an alien gene from one species into an entirely different one—has the same potential of creating an alien "monster". This monster, under normal conditions, would not have any selective advantage over native species growing in an undisturbed natural habitat. But there are few such undisturbed habitats left in the U.S. Thus, the creation of a transgenic organism and its release into the environment raise the serious threat that it could find a foothold and outcompete native species in a disturbed or manipulated habitat. The outcome of this would be unknown and the impact of the invading alien unpredictable. It is not inconceivable that huge tracts of vegetation, including food crops and their accompanying pollinators, could be wiped out and replaced by a transgenic species that itself might have no value whatsoever.

The core of evolution's lesson is adaptation. To meet changing external conditions (climate, food sources, enemies, shelter), living creatures must modify their behavior or become physically adapted. The essentially infinite diversity (number of possible genetic combinations) in animals and plants, plus the imperfections that cause genes to mutate randomly or be improperly duplicated, means that within any given population there will be individuals whose genetic makeup is particularly well suited to the existing environment. Such individuals will, under these conditions, reproduce more successfully and pass on their genome, until they are spread out in the population—natural selection at work, differential reproductive success, not violence or bloodshed. The results of natural selection are manifested not here and now but in the future—by offspring; moreover, evolution and natural selection cannot predict the future and so natural selection acts only in terms of present conditions and presently available genes.

Ω

Counter-evolutionary Technologies

Perhaps the best way to understand the impermanence and alterations of the human species is to remind ourselves of the predecessor species of modern humans. Modern man, *Homo sapiens sapiens*, has been around for about 50,000 years, 100,000 maximum. The genus *Homo* evolved from a common ancestor of humans and higher apes, chimpanzees and gorillas, and the *Homo* line split off around 5 million years ago. Many scientists would put humans and chimpanzees into the same genus (*Pan*) since we share about 98.5% of our DNA. Of course the remaining 1.5% represents the most relevant and crucial marker of humans, including language and abstract thought, thus dashing the eternally-springing hopes of those who have labored mightily to teach chimpanzees to speak.

Although new fossil evidence and theories about human origins continue to appear, generally the hominid line is this: *Homo habilis* (2.5 million to 1.8 million years ago); *Homo erectus* evolving and lasting until about 500,000 years ago, then evolving into *Homo neanderthalensis* around 250,000 years ago, and thence into *Homo sapiens* about 200,000 years ago, with modern man appearing 50,000 years ago or possibly somewhat earlier.

Neanderthalensis co-existed with *sapiens* for about 100,000 years (think about that), and then went extinct about 35,000 years ago, with the most recent studies indicating some interbreeding. The reasons given for this extinction are various, including the superior hunting skills of *sapiens* which made valuable concentrated protein a larger part of the human diet and thus increased the species' fitness as well as its cognitive abilities, including co-operation in hunting large mammals and food sharing, which in turn contributed to highly adaptive social relations such as pair bonding and extended families.

Modern humans' faculties permitted and in turn were refined by greater socialization, culture and eventually language. Humans became masters of their environment not by supplanting evolution but by joining culture to it, thus becoming a product of two forces whose separate contributions and effects are no longer distinguishable. The "nature/nurture" debate can never be resolved.

But this does not mean escape from the orbit of Nature's laws. As we manipulate our environment it becomes even more crucial to make ecologically adaptive choices. Neither technology nor ethics suffices alone; technology depends upon finite resources, and an ethical just society is not necessarily an ecological one. As of today, our technological choices have created conditions that the snail's pace of evolution cannot meet in time. The false promise of genetically altering human genes to resist technological hazards not only endangers the gene pool but puts a select few in charge of far-reaching decisions about what risks are acceptable and what should be done regarding such risks. In effect they are put in charge of developing criteria for human existence.

Such judgments are in fact being made now by government regulatory agencies such as the U.S. Environmental Protection Agency or the Nuclear Regulatory Commission, which are empowered to set pollution, health and safety standards. These are in reality simply subjective value judgments about how many illnesses or deaths are politically "acceptable" and are usually based on the ability of the regulated industry to meet such standards and still make a profit. In other words, standards are set primarily to keep industry in business. And not surprisingly, those technologies which put humans at the mercy of the criteria setters are the most dangerous—nuclear power and genetic engineering—and doubly so because the guidelines by which they must operate, even to minimize harm, inevitably conflict with our basic democratic values and institutions.

There are of course sound scientific arguments against many complex technologies. Synthetic organic chemicals, ionizing radiation and genetic manipulation are, each in its way, profoundly counter-evolutionary. Our immune system and organs evolved to combat naturally occurring threats such as bacteria, protozoans and other parasites. In time of stress, shock and accident, the human body calls on numerous defenses and can be very resilient. But never in our evolutionary history were we exposed to synthetic organic chemicals or huge doses of ionizing radiation.

As for genetic alteration, it is a well-established fact that although random gene mutations are necessary for natural selection, most mutations that do occur are deleterious, not beneficial. Naturally occurring dangerous elements do not circulate freely in the atmosphere, soil or water but are tied up in compounds or geologic formations and become accessible only with the help of technology. And life on earth became possible only when cosmic ionizing radiation was reduced to extremely low levels; today, the dispersion of artificial radionuclides through nuclear fission is reversing this trend and introducing additional random genetic mutations into the human gene pool.

Genetic manipulation of the human gene pool to "improve" the human genotype, phenotype, or behavior (eugenics) is ecologically and politically dangerous because it sidesteps natural selection by replacing the evolutionary imperative of adaptation with human requirements, and, again, because it allots ethical decision-making powers to a selected elite. Who determines which traits are desirable or undesirable? Who can say that a particular trait, now distasteful or disdained, may not prove adaptive or necessary in a different environment sometime in the future, much as the heterozygous sickle-cell trait confers immunity to malaria on Africans?

Ω

Ecosystems as Models for Human Systems

Are there lessons that we can learn from the trial and error of evolution's path that can provide any guidance for human behavior and social organization? More to the point, can ecology be the basis for social change? It is arguable that comprehension of human society can be deepened by comprehension of the myriad nonhuman species and systems with which we share the planet. What are some of these lessons?

Life on earth arose only once, and all life forms have the same single origin and chemical composition, the same genetic "words" in varied arrangements. And the same mechanisms that guide the formation of new species and the adaptation of populations (natural selection and random genetic changes) have guided human development, including the human brain, source of culture, technology and values.

Empirical observation, which quite sufficed for Charles Darwin, gives us important clues to the conditions under which successful natural systems thrive. Random change must be minimized (evolution requires these but quite rarely); external events such as climate change, vulcanism, drought, fires, floods, though often causing temporary disruption, eventually cease, life forms re-appear and systems restore themselves with a new equilibrium. Transgenic engineering is perhaps the ultimate in random change, in the sense of a single, isolated change carried out for a narrow purpose and imposed without the force of natural selection.

The most severe perturbations are often seen in simple ecosystems with few species and short food chains, and scientists have tended to view systems containing greater species diversity as the most stable and resilient ones, such as tropical rainforests, where blights and pestilences find it hard to take hold because no one species dominates and most niches are already filled. However, this is not always the case, particularly where humans initiate environmental change. In tropical rainforests, species numbers are high but the numbers of individual organisms within a species are very low. In addition, such individuals may be highly specialized or have mutualistic and co-evolutionary relationships with other species. Thus, destruction of only a few hectares of rainforest could well take entire populations of several species and their co-evolved neighbors into extinction.

Stability is achieved by internal homeostasis, such as the gaseous balance in plants, by the tendency of living things not to change as long as the environment is stable, and by the subdivision of larger ecosystems into smaller biotic communities and associations. A feature of this tendency of life towards stability is that the life processes involved are self-regulating. Sensory mechanisms detect external conditions and elicit appropriate responses; the human nervous system and our emotional responses, as well as our physiological functions, are examples.

However, the spread of technology has to a large extent (and in some cases entirely) replaced long-evolved automatic self-regulating processes. Such dependence commensurately reduces stability in both natural and human social systems, rendering them highly vulnerable in times of change. The Ecologist's *Blueprint for Survival* said:

> "To suppose that we can ensure the functioning of the ecosphere ourselves with the sole aid of technological devices, thereby dispensing with the elaborate set of self-regulating mechanisms that has taken thousands of millions of years to evolve, is an absurd piece of anthropocentric presumption that belongs to the realm of pure fantasy". (*The Ecologist*, Vol. 1, No. 1, 1972).

Because members of a species function within an ecological niche not occupied by those of another species, they are subject to natural limits on their expansion, primarily availability of food and shelter. This niche existence thus controls population size and also helps to exclude potential intruders, but when the system is artificially disrupted by, for example, pesticide spraying or nutrient runoff, holes are punched in the system that then provide a foothold for pest species, weeds, algae, *etc.* Thus, the simplification of food chains by humans for human purposes creates instability and vulnerability in natural systems, which in turn stimulates demand for additional external technological controls which again in turn create new problems and require more intervention and so on *ad infinitum.*

Preserving the diversity and complexity of ecosystems, if not human social systems, would therefore seem to be facilitated by enhancing their inherent stability and self-regulation rather than imposing short-term human needs and technological regulation upon them. Such self-regulation in the case of human systems is possible only when the scale is small enough for cooperation, interaction and comprehension. A comparison of a large city and a small village makes it clear that as population increases, so do the opportunities for random change and disorder; in social terms these then require outside and/or technological control by police, firemen, bureaucrats, lawyers, electronics, etc. Thus, self-regulation of human social systems can only be possible in small units where cooperation, altruism, human sanctions, moral codes and non-coercive non-technological control mechanisms can function.

Self-regulation in natural systems—forests, rivers, swamps, reefs, grasslands, oceans, deserts—becomes possible only if humans refrain from placing external stresses on the system, resist the urge to simplify the ecosystems or direct them for human needs, and in general permit unhindered evolutionary processes to unfold. In human (cultural) terms this means preserving community and tribal cultures, which comprise cultural diversity, rather than acceding to the cultural homogenization now occurring across the globe. Similarly, captive breeding of endangered or threatened species for future release in the wild can only be effective if these species' natural habitats remain in sufficiently healthy condition and of adequate size to enable these species to repopulate these habitats.

Through the interaction of genes and environment, living things have appeared, evolved, bred, colonized and persisted in specific habitats. Species isolated from their natural habitats literally cease to exist as viable species, even when maintained artificially in zoos or gene banks. Even seed banks intended to preserve indigenous varieties of plants are no insurance because the seeds are banked in isolation, free of environmental pressure and natural selection. And if and when environmental conditions change, banked seeds not planted regularly in the wild will not have evolved in nature to adapt to the changed conditions and stand little chance of surviving when planted.

Because of the single origin of all life, we are tied biologically, chemically, physically and ethically to all other life forms. We rely on the same energy sources and physical elements for sustenance. Moreover, the differences between human and non-human life forms are quantitative, not qualitative: brain size, organ and system development, dominance of certain sensory organs and recession of others, thumb and hand facility, organs of communication, instinct and emotion. In many respects we are inferior; we lack built-in defenses like claws and fangs, our movement is slow, our senses less acute than those of many other animals, we lack protective body covering, and so forth. But our larger, more complex brain—its complexity both a result and cause of culture—enabled us to develop language, abstract thought, culture and technology, all of which compensate for biological deficiencies. Culture, in consort with decelerated natural selection, is the main force today in human evolution.

All of this indicates the need to focus on ecology and evolution—modeling our society on Nature's rules and systems—as the political organizing principles and the paradigms of our public life. We need to make the right choices for the right reasons rather than relying on arbitrary ideologies to formulate ethical and social codes of behavior. Small wonder it is, then, that real environmental solutions, a broad-based ecological ethic and a convergence of disparate movements (animal rights, environment, feminism, labor, minorities, anti-war, social justice) have so far escaped us.

Ω

Progress and Purpose

Inevitably the discussion of evolution leads to a discussion of the meaning of progress, both in terms of evolution and of human achievement. The word implies some kind of movement, and therefore presumably a goal, but not all movement and change is beneficial in the abstract. Whether one is moving towards something "good" or "better" depends upon one's view and the criteria used. Some criteria that have been suggested for measuring progress in the evolution of life forms are: adaptability, ability to manipulate one's environment, resilience, longevity, complexity, abundance, dominance (filling or controlling many niches).

Biologist Edward O. Wilson has tried to clarify the confusion created by the word "progress", which to most minds indicates a goal; there is no such thing in our genes. Wilson points out that humans and many other organisms have survival strategies and therefore chose certain actions or goals: "*ex post facto* responses to the necessities imposed by the environment . . . life is ruled by the immediate past and the present, not by the future" (*The Diversity of Life*, pp.186-7). But he also says there is another meaning of the word if one realizes that the overall movement of life has moved "from the simple and the few to the more complex and numerous". He points out that, generally speaking, "animals as a whole evolved upwards in body size, feeding and defensive techniques, brain and behavioral complexity, social organization."

By some of these criteria, humans are the most successful (environmental manipulation, complexity, dominance), but by others (longevity, resilience, abundance) we lose out to bacteria and insects (750,000 known species and growing, versus one human species and 4000 mammal species). For those oriented to humanistic concerns, perhaps the criterion of consciousness is preferable, and even some ecologically-minded New Age leaders regard human consciousness as the pinnacle of creation, thus elevating our species to a superior position, which is of course what got us into trouble to begin with.

Perhaps it might make more sense to define the most successful species as that which recognizes the unity and equality (in the "eye of God") of all life forms and acts to preserve them, a definition that recognizes both our special intellectual endowment as well as our moral obligation to the rest of Nature. So, while we are intellectually, biologically and emotionally hominid products of biological evolution, ethically we are products of cultural evolution which should ideally dictate choices consistent not only with our own survival but with the extension of basic rights to non-human species based on both biological necessity and ethical imperatives. Such choices for us are thus both morally correct and evolutionarily adaptive.

Some people are filled with despair at the notion that the human species may not be the "purpose" of life or evolution. Even agnostics and environmentalists may find it difficult to live with the omnipresence of uncertainty, whether or not they believe in a deity. And then there are those who, though cognizant of our ties to the rest of life, see enlightened, wise humanity as the culmination of the "progression" of life, responsible for stewardship of the natural world because of our superior intelligence. At the other extreme are the technocrats and scientific priesthood who regard all of Nature as resources to be exploited, humans included.

Thankfully there are many unfathomed aspects of life on earth which neither religion nor philosophy nor science will ever explain. No one has been able to demonstrate that evolution has any goal or that any species can be termed "higher" than any other (a point stressed by Darwin), or that evolution towards conscious life forms was a necessity and by definition progress. Nor has anyone been able to devise criteria or tests for any of these beliefs, hence they remain irrational and not amenable to scientific argument.

To postulate a goal in evolution would be to endow its forces with direction, suggesting design and/or intent, i.e. the will of a deity. Believing that we, the human species, are the culmination and purpose of evolution, or that we were necessary and inevitable, are simply religious beliefs that suit our personal world view or upbringing, and help justify our behavior, including unfortunately its most counter-adaptive components. But we need not turn to a creative deity or instill a purpose in evolution in order to understand the mysteries of life. Evolution and natural selection, properly understood, are fully capable of explaining how life evolved, how it works, and why it persists. What they also tell us is that there is no dividing line between what we call human and what we call non-human but rather a continuum of life forms stretching backwards in time, and, if we are lucky, forward. Science can explain the "how" of life on earth; it is up to individuals and religion to muse on the "why".

Ω

Humanity in the Biosphere

The late Murray Bookchin raised the ethical and philosophical problems associated with a hard-line biocentric view, that is, the view that all species have an equal right to exist and evolve, including viruses and other pathogens. There are real issues involved in deciding whether and to what extent human intervention between harmful organisms and their targets or prey should be permitted. What are the criteria for extinguishing an entire species? The actual or potential harm? And there is the corollary: is there an objective way of certifying and quantifying this potential harm? If this pathogen kills or could kill one person each day? Each week? Each year? Or more people? Do they die all at once, or just get sick? Is the disease preventable or curable? What are the predictable side effects of species extinction? What are the unknowns and uncertainties? Will hardship or suffering accrue to others as a result? And while several "friendly" species have become extinct or nearly so through human actions (the Carolina Parakeet and Passenger Pigeon in this country), our efforts to wipe out insect pests, our admitted "enemies", with chemicals have failed or backfired ecologically, with disastrous consequences for human health, the environment and non-target species.

Human morality does in fact require us to prioritize, that is, frequently put human welfare before that of other species, though we have not yet been faced directly and knowingly with the thorny question of balancing the survival of the whole human species against that of an entire non-human one. What we do face continually, however, is the sacrifice of non-human individual organisms and populations—not for disease prevention, human health or survival but for expedience, convenience, amusement, or profit. This dilemma is a social and political one and as such, it justifies our taking the position that without a compelling moral argument (survival of human individuals, populations or species), sacrifice of the right of non-human organisms or species to live and evolve should not be countenanced.

Additionally, from a purely pragmatic viewpoint, the species being sacrificed, and/or its habitat, is lost forever and can never be recalled; no "alternatives" exist to this life form. But in the cases of human endeavors, there are nearly always alternatives (which may be inconvenient or costly but never life-threatening). The shutdown of a chemical factory may create unemployment but alternative employment is in principle available, provided political will, legislation, court fiat or other means are exerted.

A word on human consciousness is appropriate here. The fact that we are, or believe we are, the only self-aware species on earth (something that cannot be proven) does not mean that this was evolution's impulse or "intent" or our own "striving". We need not have survived at all; there was and is no "necessity" for *homo sapiens* to continue to exist. That we did survive, however, can be explained quite satisfactorily by studying our evolutionary ancestry. We can trace it back to intelligent tool-making hominids, or further back to very early primates with useful opposable thumbs and stereoptic vision, or even further back to mammals, whose warm blood, long internal gestation period, fur and other physical traits were extremely advantageous after the dinosaurs died out.

Consciousness, ethics and morality are not qualitatively different from many non-human behavioral traits such as care of young, defense of the tribe or flock, play, companionship, maternal love, mourning, food sharing and so forth, but only quantitatively, that is, more highly developed. In terms of natural selection, adaptability of physical and behavioral traits means persistence, refinement and broader distribution throughout the population. Higher consciousness, including all the intermediate stages of its manifestation before *homo sapiens* appeared, was not uniquely innate or "sought out" by human beings; rather, its more primitive stages were themselves adaptive enough to persist; those individuals or populations who possessed it survived and reproduced more successfully. This insured the perpetuation of those same intermediate stages of consciousness, which then made further changes possible. The accumulation of small random adaptive changes through natural selection transformed those individuals and populations repeatedly over time—in brief, evolution.

Without exception, at least in the developed world, industrial, commercial and land development schemes are not intended to produce ethically or socially desirable objectives (saving lives, preventing deaths) but are usually based on flimsy and transparent arguments about potential benefits to the community or a particular group. It is in this kind of setting that ecological ethics must be applied, not least because the social impact of such ventures is usually at least as destructive as the environmental impact. Industrialism is horrendous not only because it wipes out natural habitats and species but because it inevitably throws out of balance the social and cultural values of a particular place or community, and more often than not that community is a particularly disempowered or minority or impoverished one.

The lesson of ecological ethics and of "deep ecology", therefore, is not that human welfare and the human species are no more important than the rest of Nature but that, for all practical purposes, in our political struggles against destructive industrialism, we need to believe and behave as if they were equal—a functional equality as it were. Thus, the issues raised by economic growth, industrialization and the spread of risky technologies are not moral ones ("environmentalists care more about birds than people") but social and political ones. Cases purportedly involving birds vs. development illustrate *social* conflicts such as a desire for profit or comfort; but curing disease or destroying the smallpox virus is a matter of an ethical duty to fellow humans. These must not be confused.

Ω

The Ecological Imperative

Homo sapiens sapiens is the only species that comprehends that it is part of the biosphere. Yet despite this knowledge, humans rely on their supreme adaptability to conditions that are in the process of destroying them. It is questionable whether humans have any "survival" instinct. In *Beyond Survival*, David Dellinger wrote: "I was taught that the instinct for survival is basic to all forms of life and that no species could survive without it . . . the communal instinct for survival of the species is asserting itself in spectacular fashion. Wave after wave of public opposition to nuclear weapons sweeps over Western Europe and the United States . . ."

In fact there is no such instinct in humans; there are, however, instincts to take steps that are conducive to survival. We do not eat food and drink water because we tell ourselves we must eat and drink to survive; we eat and drink because we are hungry and thirsty. Indeed, the fact that many people profess to seek peace and act to the contrary (and here I mean sane, moral people, not criminals or the insane) or act indifferent, would seem to indicate that survivalist arguments for peace or environmental sanity are not very effective.

Life is made possible because of the adjustment between a species' requirements for survival and the physical limits of its environment. This continuing process of adjustment is evolution, and its product is biological diversity, the continued emergence of new life forms from pre-existing ones. But at the risk of extinction humans have chosen to ignore the conditions of their existence in much the same way that they ignore the conditions of their origins. In order to survive, we must understand both. This does not mean an instinct for pure survival but rather the desire to alter or improve the conditions under which we live: social progress. In this sense, social change movements could be characterized as movements for survival, but the crucial dimension is lacking—the dimension of Nature, of the biosphere in which all human societies and communities must function.

Technology has erected a wall between humans and nature resulting in severe alienation from nature, but our political institutions are equally to blame. The more remote, arbitrary and oppressive these institutions, the more alienated the citizenry from action and from the consequences of inaction. Pollution of our environment is not merely a collection of unpleasant physical phenomena that can be dispelled, concealed or diluted by large doses of scientific expertise. It is the concrete result of basic dysfunction of human-made systems, and of decisions made by discrete individuals, and as such should provoke us to re-assess not only our attitudes towards Nature but the character, scale and purpose of existing institutions, particularly those that mediate our relationship to Nature. In this sense, environmental problems should be seen not as isolated cases or accidents but rather as the indicators of an entire system—one fixated on endless production, consumption and expansion—gone horribly wrong, a system that Nature ultimately does not tolerate.

As the specter of nuclear war recedes somewhat, an equal and perhaps greater peril faces us: biotic impoverishment from the extinction of species. The only difference between these two threats lies in the time frame; nuclear war will kill off "higher" life forms first, while species destruction extinguishes "lower" forms first—and is happening right now. The risks to the cosmic genetic library contained within living organisms—all the biotic books that began with the same word—can be recognized through ethical inquiry but more completely and forcefully by an understanding of evolution and ecology.

Evolution, by placing human beings in the middle of the long continuum of life, can be a guide (though not always a criterion) not only for science and philosophy but also for culture, politics and ethics. It can enable us to address more rationally our social structures and political goals, as well as our scientific concerns like energy, water, food, resources, pollution, health, and, not least, philosophical ones (the origins of human and non-human life, the common roots of all life forms, the unfolding of ethical concern

65

for other species, the purpose of human existence). Its utility in helping us evaluate social institutions critically arises because it makes us ask the right questions, not just those that are convenient or unthreatening.

The significance of today's ecology and Green movements lies not only in their expounding of a biocentric philosophy and ethic that embeds humanity within Nature but in the potential of this philosophy for becoming an ethical and pragmatic replacement for political ideologies that deny or violate Nature or rationalize human oppression as well as for religious cults and movements intent on banishing rational thought and replacing science with superstition or personal political doctrines.

Human systems do not, by dint of their origin, escape the constraints governing the rest of Nature. Neither political rhetoric nor sophisticated technology nor scientific hocus-pocus can obscure this fact. To presume that our higher intelligence and technological achievements somehow exempt or insulate us from the workings of Nature is a blasphemous presumption of infallibility and immortality. Our talent for manipulating Nature and our ability to adapt to horrendous squalid living conditions are only temporary insulation, and for a very short period at that.

Ecology, the study of the interaction of living things with each other and with their environment, comprises social as well as biological relationships. The social and political interaction of human beings, human cultures and ethics are as important to our future as our ecological interaction with algae, whales, bacteria, tropical forests and sunlight. But it is highly unlikely that humans will assume ethical responsibility for the well-being of non-human species and the integrity of natural systems unless and until they possess the political power to control the conditions of their own daily existence in human societies. An ecological commitment presupposes a political one, which presupposes political empowerment.

There is no Utopia. Humans are frail, changeable, imperfect, irrational, insane, malicious, ignorant, unpredictable, unstable. Genetic engineering will never alter this, as evolution and the laws of genetics demonstrate. Thus, social change—institutional, systemic, political—is our only available tool. We must strive to re-structure our institutions so as to minimize the opportunities for mischief—that is, reject technological fixes in favor of those that are resistant to human failure and whose malfunctioning will have minimal social and ecological impact.

Our present industrial way of life is more than treacherous; it is treasonous. Its character and goals are by necessity destructive of those values we profess to hold dear, and equally destructive of the very foundations of life on earth. It persists only because of the suspension of the normal forces that could control or re-direct it: truly free enterprise, accountable elected officials, an uncorrupted press, socially responsible businesses, independent universities, democratic decision-making, and self-government. A growing body of Americans, not sufficiently angry to rebel, sense in their bones something awry in America. At the very least they may be spurred to seek "personal growth" (a politically evasive stance); at most they may join in a citizen movement to encourage more radical societal transformation, not through violence but through the de-legitimization of bureaucracy, transnational corporate oligarchy, and technocracy, recognizing these not as tools for social progress but impediments.

At the same time, perhaps, there will be those honest enough to recognize where traditional social change movements have fallen short: in their acceptance of the conventional terms of debate and struggle; their subservience to the two-party system; their lack of an integrated world view comprising human and non-human communities; their failure to recognize environmentalism as a social justice movement; their skepticism of science and scientists; their subsuming of ragtag political theories and doctrines under single-issue campaigns; their indulgence in superficial mass organizing; their belief in the value of continued economic growth; their tacit acceptance of industrialism and its mass culture; and most crucial, their failure to analyze critically the political roots of the global ecological crisis.

The case against the religion of industrialism and economic growth is as severe an indictment of western society as the case against apartheid, poverty, homelessness, illiteracy or political repression, for these, indeed, are its by-products. Nor is this impact limited to one class, one race or one country. It is far more deadly, more pervasive, and ultimately irreversible. In the end, the "interest group" it threatens is that assemblage we call Life. And the backlash of industry and government against ecological consciousness is the clearest proof we have of their recognition of the radical threat of ecological thought to their way of doing business. Now social change movements must come to that same recognition.

The essential ecological insight is that humans and their societies are products of Nature and in the midst of an evolutionary process in which they are, seen from outside the universe or from "the eye of God" no more or less significant than other creatures. Ecology thus can be a cornerstone not only for new ethical and spiritual beliefs but an adaptive basis for social, economic and political reform, an alternative to abstract, human-centered ideologies, a guide to the construction of a sustainable equitable society with profound spiritual underpinnings.

The implication of refusing to recognize and honor Nature's role in shaping the human species is not equity or social justice but potentially arbitrary and capricious rule; if humans need not defer to Nature, then human societies are vulnerable to control by despots and dogmas, guided not by any higher values or ethical constraints but by the imposed beliefs of those in power at the moment.

Those who are suspicious of Darwin and evolution have overlooked the fact that the grand plan of the Industrial Revolution and early natural theology were threatened, not strengthened, by evolutionary theory, since evolution, properly understood, is inherently subversive of and contradictory to abstract ideologies. While today's industrial society rationalizes control over people and resources in spite of Nature, with its own kind of social and technological determinism, evolution's lessons contradict a technocratic hierarchical view of the earth as much as they did in the nineteenth century. The lessons of evolution and natural selection are not those of hierarchy, dominance, violence and inequality but of unity, continuity, diversity, interdependence, equilibrium and sustainability. In a sense we must re-do what Darwin did for his day: crumble the foundations of anthropocentrism and rebuild an ethics of biocentrism in its place.

The apparent mystery, beauty and fascination of Nature are by themselves worthy of reverence and homage; the fact that (to paraphrase Darwin) from primitive beginnings such wonders have unfolded is as miraculous as any theistic dogma. The spiritual inspiration we gain from the endless miracles of Nature's work is surely a foundation for human values far sturdier than any organized religion, and far more capable of re-defining and resolving the major moral crises of our time through new adaptive values and structures.

Ecology is in fact the only extant idea with a future. The recognition of the common origins and interdependence of life forms within the biosphere is perhaps the highest moral awareness of which humans are capable. The challenge to global citizens is to create a new politics out of this awareness.

Ω

Formulating a New Agenda for Social Change: Nature Comes First; Nature Bats Last

The conditions of human existence and society are dependent on the preservation and integrity of the world's natural systems. Without instilling this basic truth as the philosophical and political center of a movement for social change, goals are muddied and action deflected elsewhere. Traditional religions were historically at the center of human society and offered their own incorrect explanations for how the world worked. Now we live in the post-Enlightenment world where science provides explanations that are reliable, pragmatic and useful. But they are also imperative to the decision- and policy-making bodies. For these to function honestly and with accountability, the citizenry must itself be educated on the ecological exigencies of living on earth.

Any campaign or movement for change must therefore be responsible for bringing an ecological consciousness and conscience to the public. Actions arising out of the movement must be explicitly connected to remedies for the most pressing ecological problems and averting the gravest threats. Today these are climate change, loss of biodiversity, overpopulation and destruction of the oceans. Within each of these there are subcategories and pressure points, in the fields of energy policy, land use, depletion of freshwater resources, overharvesting and pollution.

From Marxists and libertarians, a frequent assumption appears: that a correct economic theory—one which forces producers to include the externalities of production in the price of their products—will make everything else in society fall into place. Both free marketeers as well as Marxists thus believe in "economism", i.e. that economic reform and adjustment will bring about reforms in other spheres and sectors. The Marxists speak of "economic relations" and the right maintains its faith-based ideology that regards markets as definitive.

While internalization of full costs, often called Full Cost Pricing, is imperative, this will not automatically bring about the full spectrum of political and social change that human society needs to survive. In any case economic theory should not take precedence over other responses, important as Full Cost Pricing is. Most economists today do not seem to agree; not too long ago economist Nouriel Roubini gave a talk on the economy and markets and never mentioned energy. This is quite typical of economists.

An economic system is entirely human-made, springing fully formed (or half-cocked) from the human mind. Humans can devise any system they want, like printing paper money or checks to reflect fiscal reserves or debts, but it has no external, objective or scientific basis. Economic and financial systems are simply mechanisms to serve other ends; they have no intrinsic value. Such mechanisms will behave in good or bad ways, but when they do falter or behave badly, humans will once again rely on their ingenuity to try and fix things by printing money or arbitrarily adjusting values.

As a result, any given economic system is arbitrary. This means it is not a foregone conclusion nor the only one conceivable. It lasts as long as people abide by its rules and accept the consequences. *People* create the system. They can change or abolish it. In order for any economic system, whether that using paper money or barter, to be legitimate, it has to be accepted by most of society. This means that it is preceded by a *political* system. An economic system can't exist prior to a political system because a political system must be in place before any human-constructed arrangement can come into being. Social systems such as barter-dependent communities are simply sub-sets of political systems.

So politics precedes economics. And no change in the economic system can be forthcoming unless and until the political system allows it. An economic system is not autonomous. It needs a political framework to operate.

But there is another system which precedes and underpins both of these: the biological system, or rather the ecological framework from which humans and their societies emerge and to which they must conform by adaptation or the formulation of certain rules. The biosphere is the ultimate external, non-arbitrary constraint on human behavior and on human social arrangements, technology notwithstanding. If humans resist adaptation and instead degrade or shred the components of the biosphere, their societies will surely be destroyed sooner or later, no matter how legitimate or democratic those societies are, and no matter how flexible its economic system. Jared Diamond, in *Collapse*, has written eloquently of numerous societies that extinguished themselves.

Thus, the only true exigencies that exist are those laid down by Nature and the biosphere. These are fixed and the consequences of defying them are dangerous, often lethal. It has taken post-Enlightenment society hundreds of years to learn Nature's lessons. And what we now know is that if our political and economic systems do not understand the exigencies of nature and do not adapt and conform to them, those man-made systems will collapse.

So in constructing any economic system, we need legitimacy and acceptance. These can only be found through political means, hopefully through peaceful, equitable and democratic means. But that political system, even if legitimate and democratic, is not then free to defy or ignore the constraints of nature even with the most magnanimous and humane intent. To survive and persist, it must accommodate itself to the ultimate reality: the need to preserve the integrity and evolutionary destiny of the earth's species and ecosystems.

Accordingly, the economic system must also accept this reality, and instead of relying on arbitrary, human-derived theories or philosophies, it must take its lead from the lessons of nature regarding such things as ecological requirements, interdependence, symbiosis, appropriate scale, biotic/social communities, control of population, sustainable production and consumption, recycling and detoxification of all wastes, and special attention to the global commons of air, oceans, water, soils, and biomes. In this regard it must, out of purely self-interest, imitate as closely as possible natural systems that seek equilibrium in order to maintain their integrity and reproductive future. Our present-day system is badly out of equilibrium, neither sustainable nor equitable, and arguably on the verge of either social collapse or totalitarianism.

The lesson that we need to take away from this is that economics, even the most progressive and equitable, can neither design nor precede an ecological society. If we want a sound economic system, we must first develop an ecological paradigm and theory, within which our political and economic systems must fit. We need to remind ourselves of it again and again. Economic theory follows democracy, which follows ecology. Any other order would mean tyranny and extinction.

Ω

Afterword
Why Science and Religion
are Incompatible

Science and Religion are indeed incompatible, notwithstanding statements to the contrary by some scientists and theologians.

Religion and science both offer explanations for why life and the universe exist. Science relies on testable empirical evidence and observation. Religion relies on subjective belief in a creator. Only one explanation is correct. The other must be discarded.

Religion does offer other things as inducement to believe its explanation: moral codes, rewards and punishments, personal guidance and a refuge from human pain and conflict. Were religion to abandon its belief in a creator, it would then remain as a moral philosophy with which science would not take issue. In fact some religions do come closer to philosophy; Buddhism is one example.

Unfortunately historic fanatics like Mohammed and institutions such as the Catholic papacy took advantage of human frailties, fears and innate irrationality, to create power structures and doctrinal controls over all aspects of human life, based on the mandates of a creator. For obvious reasons, it became imperative to denigrate and destroy those ideas that contradicted or undermined their doctrine and threatened their control. To defend their power, they had to insist that the doctrine was given to them by a creator, thus crushing any arbitrary secularism that might rear its head. Those who raised doubts were eliminated (and in the case of Islam still are).

Evolution was and is the most powerful challenge to religion in that it reveals the material origin of human beings, including their mind. The power of theology was immediately diminished when Darwin showed that life on earth did not require a creator. But the powerful do not readily relinquish power, as radical Islam and the stubbornness of the Catholic Church demonstrate. Nor do even the educated faithful, scientists included, seem willing to redefine and rename religion as moral philosophy.

Scientists in the laboratory must in effect shelve their belief in a creator in order to do their research. Some, such as Francis Collins, live a kind of lie in their professional work, accepting the proofs of physics, chemistry and biology but pretending that the existence of a creator has not been disproved . . . a pretense that by itself is completely irrational and unscientific.

How do atheists and believers differ? Atheists admit they cannot disprove the existence of a god. Believers refuse to admit that they cannot *prove* it.

Explanations require evidence. No evidence for a creator exists outside the human mind, whereas the evidence for evolution and the origins of life mounts every day. In the face of this uncontradicted evidence, religious belief in a divinity is no more viable than belief in the Flying Spaghetti Monster.

Ω

Recommended Reading

Evolution

Why Evolution is True. Jerry A. Coyne. Viking (Penguin), 2009.

The Blind Watchmaker. Richard Dawkins. W.W. Norton, 1987. *

The Origin of Species (illus.). Charles Darwin. Abridged by Richard Leakey. Faber & Faber, 1979.

Darwin's Dangerous Idea. Daniel C. Dennett. Simon & Schuster, 1995.

Voyaging—Charles Darwin: A Biography (Vol. 1, 1995); *Charles Darwin: The Power of Place* (Vol. 2, 2003). Janet Browne. Alfred A. Knopf, 1995.

Science on Trial. Douglas Futuyma. Pantheon Books, 1982. *

The Darwinian Revolution. Michael Ruse. University of Chicago Press, 1979.

The Eclipse of Darwinism. Peter J. Bowler. Johns Hopkins University Press, 1983.

Evolution & the Myth of Creationism. Tim Berra. Stanford University Press, 1990.

One Long Argument. Ernst Mayr. Penguin Books, 1991. *

What Evolution Is. Ernst Mayr. Basic Books, 2001.*

Beak of the Finch. Jonathan Weiner, Alfred A. Knopf 1994

Ecology

Nature's Services. Gretchen Daily, ed. Island Press, 1977.

The Machinery of Nature. Paul Ehrlich. Touchstone Press, 1986.

The Diversity of Life. Edward O. Wilson. Belknap Press, 1991.

Extinction. Edward O. Wilson. Random House, 1981.

Biodiversity. E. O. Wilson, ed. National Academy Press, 1988.

Sociobiology

Sociobiology: The New Synthesis. Edward O. Wilson, Harvard University
   Press, 2000.
The Blank Slate: The Modern Denial of Human Nature. Steven Pinker,
   Viking Press, 2002.

Other reading

Betrayal of Science and Reason. Anne Ehrlich and Paul Ehrlich, Island
   Press, 1996.
Deep Ecology for the 21st Century. ed. George Sessions, Shambala Press, 1995.
In the Absence of the Sacred. Jerry Mander, Sierra Club Books, 1991.
Sustaining the Earth. John Young, Harvard University Press, 1990.
The American Conservation Movement: John Muir and his Legacy. Stephen
   Fox, University of Wisconsin, 1985.
The Star Thrower. Loren Eiseley, Harcourt Brace Jovanovich, 1978.
Collapse. Jared Diamond, Penguin Press, 2005.
The Bridge at the Edge of the World. James Gustave Speth, Yale University
   Press, 2008.
Plan B 4.0: Mobilizing to Save Civilization. Lester R. Brown, Earth Policy
   Institute, 2009.
Dwellers in the Land. Kirkpatrick Sale, University of Georgia, 2000.
Blueprint for Survival. Ed. Edward Goldsmith, The Ecologist, 1972.
The Flight from Science and Reason. Paul R. Gross, Norman Levitt, Martin
   W. Lewis, Annals of New York Academy of Sciences 1997.
Fashionable Nonsense: Postmodern Intellectuals' Abuse of Science. Alan D.
   Sokal and Jean Bricmont, MacMillan Books, 1999.

Web Sites on Evolution

Richard Dawkins: www.richarddawkins.net
PZ Myers: scienceblogs.com/pharyngula
Aaron Ra: www.LocoLobo.org

---

*   The best introductory reading.